Disruptive Technology and Defence Innovation Ecosystems

**Innovation in Engineering and Technology Set**

coordinated by
Dimitri Uzunidis

Volume 5

# Disruptive Technology and Defence Innovation Ecosystems

*Edited by*

Pierre Barbaroux

WILEY

First published 2019 in Great Britain and the United States by ISTE Ltd and John Wiley & Sons, Inc.



ISTE Ltd
27-37 St George's Road
London SW19 4EU
UK

www.iste.co.uk

John Wiley & Sons, Inc.
111 River Street
Hoboken, NJ 07030
USA

www.wiley.com



Library of Congress Control Number: 2019936669

British Library Cataloguing-in-Publication Data
A CIP record for this book is available from the British Library
ISBN 978-1-78630-449-0

# Contents

# Introduction

## Disruptive Technology and Defence Innovation Ecosystems

### I.1. Objectives of the book

Launched in March 2018 by the French Ministry of Defence, the creation of the *Agence de l'innovation de défense* (Defence Innovation Agency) highlights the importance of innovation and research into the development of defence industrial and technological capabilities. The *Direction Générale de l'Armement*, DGA (French Procurement Agency) has launched several programs to support the Ministry of Defence's innovation policy by mobilizing large corporations, research laboratories, start-ups, small and medium enterprises (SMEs) and small and medium industries (SMIs). The objective is to promote industrial and technological skills in the fields of complex data processing (e.g. data fusion, analysis and dissemination, task automation), human–system collaboration (e.g. cooperation in teams composed of human and artificial agents) and virtual or augmented reality, at the service of military capability development (e.g. planning, command and control of operations, intelligence, surveillance and reconnaissance, logistics, education and training).

Besides this French example, innovation defines a global strategy of capacity adaptation and change management for defence stakeholders whose primary purpose is to enable the armed forces to have effective organizations, equipment and materials at their disposal with regard to military affairs. While the military model of innovation management

---

Introduction written by Pierre BARBAROUX.

remains deeply rooted in the experimental approach due to its collaboration between academic, industrial and military operators, the defence community is currently facing several challenges. Located at the forefront of several major transformations affecting its business models and skills, artificial intelligence (AI) perfectly embodies these challenges. Hence, it is a central topic for analysts and practitioners, and leads to policy and strategy discussions all over the world, particularly in Europe, the United States and China [KAN 18]. While it is not the only knowledge domain that can transform military organizations and capabilities in the long term[1], AI raises a number of fundamental issues that question the participation of defence stakeholders in the innovation process and their ability to benefit from it in operational, political and economic terms.

Advances in the disciplines of computer science and artificial sciences (e.g. robotics, virtual or augmented realities, big data analysis) are changing the knowledge structures from which military technological trajectories develop in a sustainable way. Because it develops outside the traditional defence industrial and technological base framework (DITB), AI perfectly illustrates the disruptive potential of a knowledge application field developed on the periphery of this "ecosystem" and widely affecting its various layers (e.g. system integrators, suppliers and subcontractors, research laboratories, training organizations, government agencies, civil and military operators). Embodying a dual technological and organizational disruption for defence actors, AI profoundly redefines the way in which the processes of knowledge generation, application and exploitation are organized [BAR 16]. It also justifies taking a fresh look at the uses and methods of developing users' skills [BAR 18]. Finally, it leads to the rethinking of military operations in the light of the changes brought about by the interaction between two forms of intelligence, human and artificial, which is carefully incorporated into the capabilities, practices and artifacts used to plan, command and control operations (e.g. intelligence, surveillance, reconnaissance, communication, decision support [BAR 17a]).

The objective of this book is to address the hot topic of renewing the cognitive and organizational foundations of innovation within the defence sector. It aims to discuss the implications for its stakeholders, in particular the companies responsible for creating innovations and those in charge of

1 Quantum technologies, biotechnologies, nanotechnologies and ecological technologies have an equally great impact.

implementing them. Two complementary dimensions of the transformation induced by innovation for defence are highlighted: industrial and intra-organizational. On an industrial scale, there is indeed an impact on the ecology of knowledge underlying the industrial and technological military ecosystem. Knowledge bases are changing and new methods are emerging, the boundaries of science and technology are shifting, opening up opportunities for value creation on the periphery of the traditional DITB ecosystem. How are organizations adapting to these changes? How does this change their core competencies and knowledge bases? At the intra-organizational level, it is the stakeholders' habits that change and, in turn, their skills and the processes by which they are acquired and transferred. In this context, the sources of innovation value and its legitimacy are being transformed, requiring in return sometimes radical adaptations to the role models, organizational structures and strategies of established stakeholders (companies, military, researchers, government agencies [BAR 17b]).

We have also chosen to highlight the effects of AI on stakeholder skills (organizations and individuals) insofar as it refers to a cross-disciplinary field of knowledge that is recognized as penetrating most technological innovations that may affect current and future military capabilities [GON 18]. More specifically, AI modifies two fundamental capabilities: the implementation of weapon systems and the command and control ($C^2$) of operations. The concepts of autonomy and collective intelligence applied to weapon systems and $C^2$ systems, respectively, embody two major innovations for defence organizations that build on the progress made in the field of AI.

While they are at the heart of strategic thinking within the defence community, these concepts remain poorly defined and their implications poorly perceived. Can a remotely piloted aircraft system (RPAS) be described as autonomous? What is the meaning of an "intelligent" $C^2$ system? What are the forces driving the development of innovative ecosystems, such as remotely piloted aircraft systems? More broadly, what types of knowledge support innovations are found within the defence sector? What implications do digitization and AI have for operators' skills and even for their professional identities? These questions are not clear-cut and this observation justifies further assessment of the meaning given to innovations driven by advances in the field of AI, in particular so-called "autonomous" systems and "intelligent" digital decision-making artifacts. These two socio-technical objects remain largely “unthought of” and several chapters of this book are devoted to them.

This book is organized into two parts and nine chapters.

The first part is dedicated to the transformation of the innovation organization model in the defence sector, through the study of knowledge bases distributed within this ecosystem (inter-organizational scale). It is composed of five chapters.

In Chapter 1, Jean Belin and Marianne Guille analyze the dynamics that have affected research and development (R&D) and innovation in the defence sector since the 1980s in order to understand the changes in the behavior of the various stakeholders (public or private companies, universities and government agencies). The French innovation system has undergone profound changes to which the various stakeholders have had to adapt, particularly in the defence sector.

In Chapter 2, Cécile Fauconnet explores the evolution of the innovation model in the aerospace and defence (A&D) sector between 1945 and 2015. In this chapter, the author examines the evolution of the innovation model of the A&D sector and argues that it is marked by the intensive use of scientific knowledge.

In Chapter 3, François-Xavier Meunier develops the research field introduced in Chapter 1 and proposes a method for empirically identifying the knowledge systems that structure the “technological landscape” within defence industries. The author thus identifies technological knowledge systems (TKS) from which he highlights the knowledge bricks (component knowledge) as well as the knowledge architectures (architectural knowledge) which, for each system, structure the dynamics of innovation in the defence sector.

In Chapter 4, Cécile Fauconnet, Didier Lebert, Célia Zyla and Sylvain Moura study the organization of R&D in A&D companies as well as their innovations through a “patent” approach. The authors propose a typology of these companies according to their ability to generate technological synergies in their innovation activity and according to the trade-offs they make between exploration innovations and exploitation innovations.

In Chapter 5, Pierre Barbaroux shows, through the case of remotely piloted aircraft systems (RPAS), how the development of innovations in the defence sectors is conditioned by their legitimacy, which depends on the coherence of political and technological choices made by public and private organizations (communication, innovation, regulation).

The second part of this book explores the transformation of skills and uses induced by innovations, with an emphasis on those related to the application of advances in digitization, AI and autonomous systems. It is composed of four chapters.

In Chapter 6, Vincent Ferrari questions the notion of artificial intelligence through the study of the concept of collaboration between a human agent and an artificial decision support system. Starting from a critique of the paradigm of "specialized" expert systems (i.e. a paradigm based on the application of learning algorithms to big data), the author shows that human decisions are characterized above all by the implementation of procedures that involve the application of a general principle, a rule of common sense or approximations that very often prove to be judicious: heuristics. In doing so, the author questions the foundations of "intelligent" collaboration between human and artificial agents, a collaboration that underlies disruptive innovations in the fields of command and control of military operations in a digital environment.

In Chapter 7, Nicolas Hué, Walter Arnaud and Christophe Grandemange analyze how technological innovation (e.g. AI, big data, augmented virtual reality, 3D printing, remote maintenance) is changing the practices of defence stakeholders responsible for maintaining equipment and materials in operational condition (MCO) as part of weapons programs.

In Chapter 8, Cyril Camachon and Pierre Barbaroux discuss the implications of digital innovations for military operators through the study of the changes brought about by the introduction of new generation aircraft in the training of French Air Force pilots. These new aircrafts (Cirrus SR20 and Pilatus PC-21), equipped with innovative capabilities (glass cockpits and on-board simulation), represent a technological breakthrough that could affect the nature and diversity of pilots' skills, as well as the training process governing their acquisition.

In Chapter 9, Bertrand Kirsch and Olivier Montagnier study the technological innovation represented by the high-altitude solar drone with almost unlimited endurance through the persistent technological obstacles that today limit its development and its civil and military applications. The low propulsive power extracted from the solar source imposes a very particular system architecture, with, in particular, a very long and flexible wing, particularly vulnerable to destructive interactions. The authors then show how a transdisciplinary approach, established in experimental research, offers innovative solutions to overcome the technological obstacles that hinder the development of a dual innovation.

The book's Conclusion recalls the innovation challenges faced by the defence sector. If, today, the potential productivity gains brought by AI and its applications have prompted significant investments by defence stakeholders to adapt military capabilities through the integration of new technological bricks into existing networks and platforms, the latter will eventually be replaced by combat networks, composed of intelligent, collaborative and autonomous vectors, effectors and sensors that connect human and artificial agents. In the absence of a deliberate strategy to anticipate the effects of future changes and develop capacities for innovation and immersion, the conclusion states that traditional defence stakeholders may be affected by change rather than leading it.

## I.2. References

[BAR 16] BARBAROUX, P., ATTOUR, A., SCHENK, E., *Knowledge Management and Innovation: Interaction, Collaboration, Openness*, ISTE Ltd, London and John Wiley and Sons, New York, 2016.

[BAR 17a] BARBAROUX, P., Cyber Résilience: Une capacité des organisations aérospatiales et de défense, Leçon Inaugurale de la Chaire "Cyber-Résilience Aérospatiale" de l'armée de l'air, Ecole de l'air, Salon de Provence, delivered publicly on Tuesday 05 December, 2017.

[BAR 17b] BARBAROUX, P., Apprendre et Innover: Une exploration des modalités d'adaptation et de conduite du changement économique et organisationnel, HDR dissertation, Pôle Européen de Gestion et d'Economie, University of Strasbourg, publicly defended on 21 June, 2017.

[BAR 18] BARBAROUX, P., "Learning in flow: Immersive practices and simulation-based training in the French Army Light Aviation", *CReA Working Paper Series*, 37 pages, 2018.

[GON 18] GONS, E., KETZNER, L., CARSON, B., PEDDICARD, T., MALLORY, G., "How AI and robotics will disrupt the defence industry", *The Boston Consulting Group*, p. 6, April 2018. Available at: https://www.bcg.com/publications/2018/how-ai-robotics-will-disrupt-defence-industry.aspx.

[KAN 18] KANIA, E., "New frontiers of Chinese defence innovation: Artificial Intelligence and Quantum technologies", *STIC Research Briefs, Series*, p. 6, May 2018. Available at: https://escholarship.org/uc/item/66n8s5br.

PART 1

# Transformation of the Innovation Organization Model in the Defence Sector

# 1

# Innovation Dynamics in Defence Industries

ABSTRACT. This chapter aims to analyze the dynamics of innovation specific to defence industries to understand the changes in the behavior of the different actors involved since the late 1980s. The structural changes inside national systems of innovation have particularly affected the innovation environment of defence firms. In addition to the intensification of research and development (R&D), of civil innovation, role of private enterprises, competition, acceleration of process innovation and increasing complexity of knowledge, a major reform of public R&D funding has occurred in France, which has strongly reduced direct and defence funding. Defence firms have adapted to these major changes by developing a dynamic of openness, accompanied by the Ministry of the Armed Forces, to the civil sector and other regions of the world: increase of dual technologies, of R&D outsourcing, European cooperation or relationship with universities.

## 1.1. Introduction

The purpose of this chapter is to analyze the dynamics affecting innovation in the defence sector in order to understand the changes in the behavior of the various stakeholders involved (public or private companies, universities and government agencies).

The defence innovation system has undergone profound changes to which the various stakeholders have had to adapt. These changes mainly reflect the structural changes that have affected national innovation systems (NISs)[1]

---

Chapter written by Jean BELIN and Marianne GUILLE.

1 In a broad sense, NIS includes "the economic structure and the institutional organisation affecting learning as well as research and exploration – the production system, the commercial system and the financial system are presented as subsystems in which learning takes place" (Lundvall, 1992: 12).

since the late 1980s (Mowery 2012) and have particularly affected the innovation environment of defence companies in France and in other arms-producing countries.

Research and development (R&D) and civil innovation have increased significantly, the role of private companies has increased and patenting has accelerated. The mobilized knowledge has become more complex and competition has intensified. In addition to these structural changes, which have affected all OECD countries to varying degrees, France has undergone a major reform of its public R&D financing system. The gradual decline in direct public funding from the early 1990s, particularly defence funding, was followed by a sharp increase in indirect funding in the form of tax incentives, as in other European countries, but to a greater extent.

To adapt, the stakeholders involved in defence innovation have accelerated the development of dual technologies. The defence innovation system has opened up to civil and foreign markets. Outsourcing of R&D is more prevalent and cooperation, especially at the European level, is becoming more important. The French Ministry of Defence, which became the Ministry of the Armed Forces in 2017, has also reformed its innovation management and financing system by creating an agency (Innovation Agency), funds (DefInvest) or by developing its links with universities.

We analyze these developments by first focusing on the changing innovation environment of defence industries, before highlighting in a second section, the dynamics of opening up the defence sector.

## 1.2. Transformation of the defence industry's innovation environment

The changing innovation environment of defence industries can be characterized by three main elements since the early 1980s. Indeed, defence companies have had to adapt, on the one hand, to changes in the science and technology system and, on the other hand, to increased competition and to the increasing complexity of knowledge. Finally, they have been particularly affected by the evolution of the R&D financial system, which has resulted in a reduction in their dependence on defence financing.

### 1.2.1. *Changes in the science and technology system*

The science and technology system in developed countries has undergone major changes since the early 1980s. These changes accelerated from the 1990s onwards with the gradual shift of the industrial world towards a knowledge-based economy, with the knowledge-intensive activities historically developed within the sectors specializing in information processing, and identified in particular by Machlup (1962), gradually being shared with the rest of the economy. Among these changes, the development of civil R&D, the increased role of private stakeholders, partly replacing the State, and the acceleration of patenting have particularly affected defence companies.

#### 1.2.1.1. *Development of civil R&D*

The development of R&D is one of the drivers responsible for the shift towards the knowledge economy, which also results from sustained investment in education, training and new information and communication technologies. Indeed, the growth of knowledge-intensive activities or knowledge industries is measured by combining indicators relating to knowledge production and management, such as R&D expenditure, the employment rate of graduate workers or the intensity of the use of new information technologies. Estimated at 29% of GDP in the USA in 1958 by Machlup (1962), the knowledge industry, which according to the OECD, includes high- and medium–high technology manufacturing industries, community, social and personal services and banking, insurance and other business services, already accounted for more than 50% of GDP in the OECD area at the end of the 1990s, compared with 45% in 1985, and is growing faster than GDP in most countries.

This strong growth was driven in particular by total R&D investment, which has risen sharply in most developed countries since the early 1980s (OECD data)[2]. This is particularly the case in the USA and the European Union, with the exception of the UK (Figure 1.1). We have thus gone from an overall investment, measured by the gross domestic expenditures on R&D (GERD), of 1.87% of GDP in 1981 to 2.25% in 2016 for France[3].

2 OECD (2018).

3 The Gross Domestic Expenditures on Research and Development (GERD) represents the total aggregate expenditures on R&D performed on the national territory.

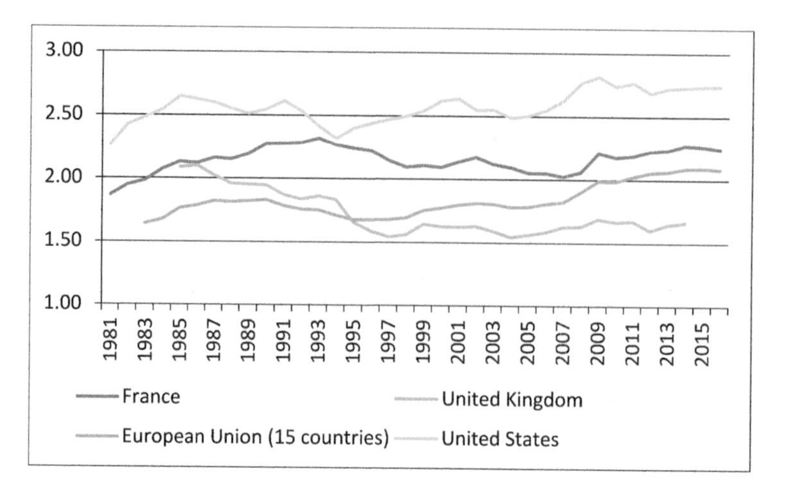


**Figure 1.1.** *Evolution of GERD as a percentage of GDP. For a color version of the figures in this chapter see www.iste.co.uk/barbaroux/technology.zip*

This evolution is explained by the sustained growth of civil R&D efforts. Indeed, in these two major areas, civil R&D effort (civil GERD to GDP) increased significantly over this period (Figure 1.2) while defence R&D effort (measured as the difference between total GERD and civil GERD) decreased as a percentage of GDP (Figure 1.3). Thus, defence GERD, which represented 0.37% of GDP in France in 1981, was only 0.07% in 2014 (the latest year available).

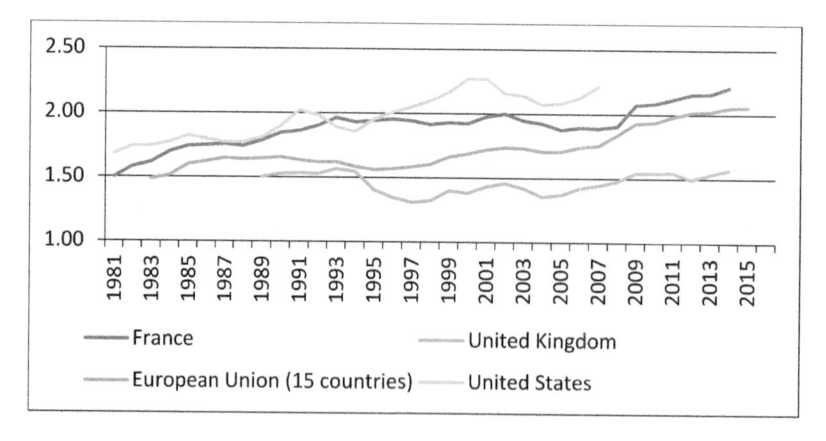


**Figure 1.2.** *Evolution of civil GERD as a percentage of GDP*

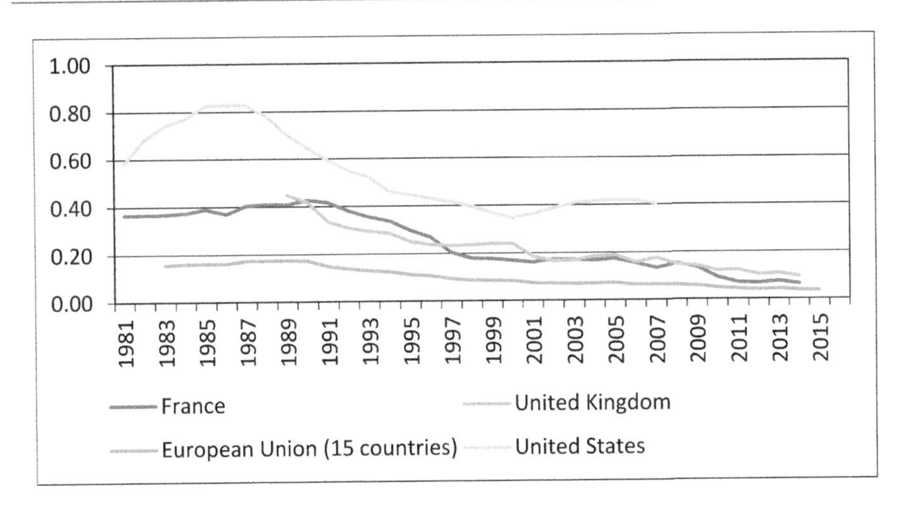


**Figure 1.3.** *Evolution of defence GERD as a percentage of GDP*

### 1.2.1.2. *Increased role of companies and universities, which are lagging behind for the State*

The trend towards an increase in civil R&D has been accompanied by a strengthening of the role of private stakeholders over this same period. Indeed, R&D activity by companies or universities is increasing, while the role of the State is tending to diminish (Table 1.1). Thus, in France, the share of companies in the execution of R&D increased from 58.9% to 63.6% between 1981 and 2016, while that of universities increased from 16.4% to 22%. At the same time, the share of the State is decreasing significantly, from 23.6% to 12.9%. Similar trends can be found in the European Union and the USA. The UK is the country where the decline in State execution of R&D is most pronounced.

| | | 1981 | 1991 | 2001 | 2011 | 2016 |
|---|---|---|---|---|---|---|
| France | Ent. | 58.9% | 61.5% | 63.2% | 64.0% | 63.6% |
| | Univ. | 16.4% | 15.1% | 18.9% | 20.9% | 22.0% |
| | State | 23.6% | 22.7% | 16.5% | 13.9% | 12.9% |
| | NPI | 1.1% | 0.8% | 1.4% | 1.2% | 1.6% |
| UK | Ent. | 63.0% | 67.1% | 65.5% | 63.6% | 67.0% |
| | Univ. | 13.6% | 16.7% | 22.7% | 26.0% | 24.6% |
| | State | 20.6% | 14.5% | 10.0% | 8.6% | 6.3% |
| | NPI | 2.8% | 1.8% | 1.8% | 1.8% | 2.1% |

| | | 1981 | 1991 | 2001 | 2011 | 2016 |
|---|---|---|---|---|---|---|
| EU-15 | Ent. | 62.4% | 63.5% | 64.4% | 63.6% | 64.9% |
| | Univ. | 17.6% | 18.8% | 21.7% | 23.3% | 22.9% |
| | State | 18.6% | 16.8% | 13.0% | 12.0% | 11.2% |
| | NPI | 1.3% | 0.8% | 0.9% | 1.1% | 1.1% |
| USA | Ent. | 69.3% | 71.1% | 72.1% | 68.4% | 71.2% |
| | Univ. | 9.7% | 11.3% | 12.0% | 14.5% | 13.2% |
| | State | 18.5% | 14.8% | 11.9% | 12.8% | 11.5% |
| | NPI | 2.5% | 2.9% | 4.0% | 4.3% | 4.1% |

**Table 1.1.** *Evolution of R&D execution*

Over a longer period (1953–2008) with American data, Hirschey *et al.* (2012) show that these two trends are clear: an increase in R&D spending and a shift in R&D spending from the government to the business sector (Figure 1.4).

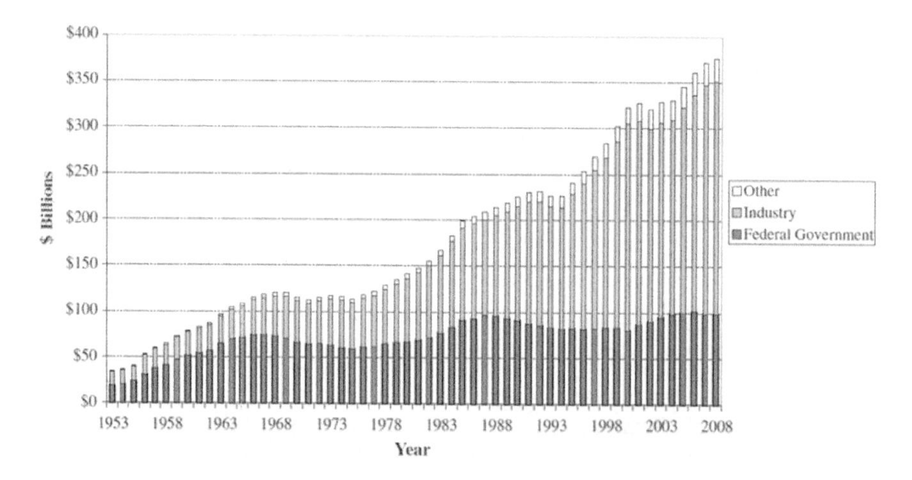


**Figure 1.4.** *Evolution of R&D execution in the USA (Hirschey et al. 2012)*

### 1.2.1.3. *Acceleration of patent filing*

The expected output of the R&D process is innovation. It is difficult to measure; some innovations are not patented and companies prefer to keep them secret. Nevertheless, as the OECD points out, "patent data can be considered as a measure approximating R&D results translated into inventions".

An analysis of the total number of triadic patent families[4] filed shows that this indicator increased in all areas studied between 1985 and 2016 (Figure 1.5). However, this upward trend is much stronger in the USA than in the European Union (Table 1.2), where French growth is above average, unlike that of the UK (see also Park (2008) and Papageorgiadis *et al.* (2014) for an international comparison).

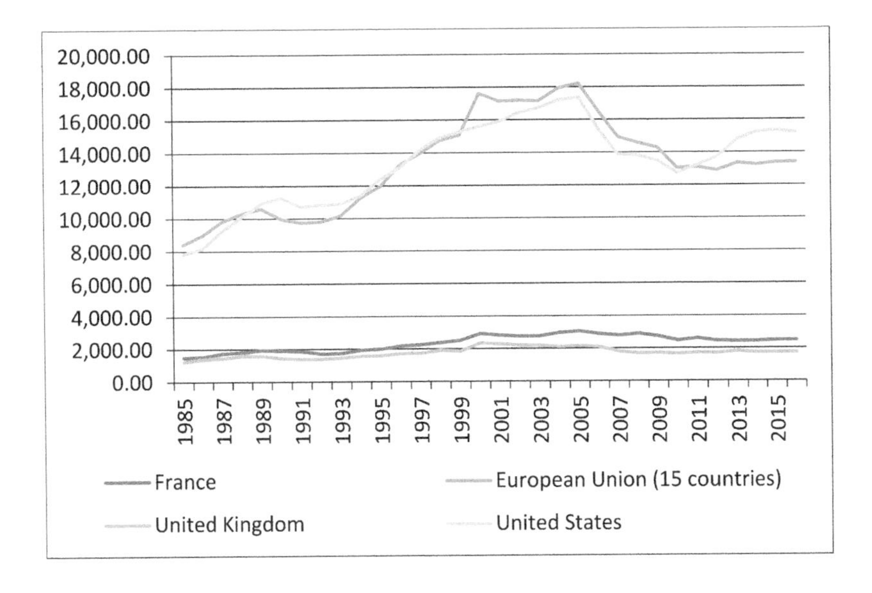


**Figure 1.5.** *Number of triadic patent families (priority year)*

4 "A triadic patent family refers to a set of patents filed in several countries (i.e. patent offices) to protect the same invention. Triadic patent families are a set of patents filed with three of the main offices, namely the European Patent Office (EPO), the Japanese Patent Office (JPO) and the United States Patent and Trademark Office (USPTO). The triadic patent family counts are attributed to the inventor's country of residence and refer to the date on which the patent was first registered. This indicator is expressed in volume". Source OECD.

| | 1985 | 1991 | 2001 | 2011 | 2016 | Growth 1985–2016 |
|---|---|---|---|---|---|---|
| France | 1,494 | 1,864 | 2,809 | 2,413 | 2,470 | 65.3% |
| UK | 1,265 | 1,406 | 2,290 | 1,818 | 1,740 | 37.6% |
| Europe | 8,419 | 9,756 | 17,159 | 13,350 | 13,402 | 59.2% |
| USA | 7,828 | 10,722 | 15,902 | 14,781 | 15,219 | 94.4% |

**Table 1.2.** *Number of triadic patent families (priority year)*

### 1.2.2. *Intensifying competition and increasing complexity of the knowledge mobilized*

At the same time, since the late 1990s, defence companies in the main armament-producing OECD countries have had to cope with both a sharp increase in competition and the increasing complexity of the knowledge and research activities used to produce the various weapon systems.

#### 1.2.2.1. *Intensifying competition*

French, English and American defence companies have gradually had to face greater competition. This competition comes from the emergence or development of the defence technological and industrial base (DTIB) from other countries as well as from civil companies that have invested heavily in R&D.

Globally, the intensification of competition can first be demonstrated by analyzing the evolution of the arms market shares in the various producing countries. This analysis can be based on the SIPRI ranking of the world's leading defence companies between 2002 and 2016 (Table 1.3).

It appears that countries such as Russia (+6.5%), South Korea (+1.8%), Turkey (+0.6%) and India (+0.6%), and more marginally Australia, Israel, Poland, Ukraine, Brazil or Sweden, have increased their market share, while France (–2.1%), the UK (–2.2%) or more significantly the USA (–4.1%) have reduced theirs.

| | 2002 | | 2016 | |
|---|---|---|---|---|
| | Market share | Number of companies | Market share | Number of companies |
| Australia | 0.2% | 1 | 0.6% | 3 |
| Brazil | 0.0% | 0 | 0.2% | 1 |
| Canada | 0.6% | 2 | 0.2% | 1 |
| France | 6.7% | 9 | 4.6% | 6 |
| Germany | 3.3% | 8 | 1.5% | 3 |
| India | 1.0% | 3 | 1.5% | 4 |
| Israel | 1.6% | 5 | 1.9% | 3 |
| Italy | 2.8% | 5 | 2.5% | 2 |
| Japan | 2.6% | 6 | 2.0% | 5 |
| Norway | 0.2% | 1 | 0.2% | 1 |
| Poland | 0.0% | 0 | 0.3% | 1 |
| Russia | 1.0% | 4 | 7.6% | 13 |
| Singapore | 0.4% | 1 | 0.4% | 1 |
| South Africa | 0.1% | 1 | 0.0% | 0 |
| South Korea | 0.5% | 2 | 2.3% | 8 |
| Spain | 0.3% | 2 | 0.2% | 1 |
| Sweden | 0.6% | 1 | 0.7% | 1 |
| Switzerland | 0.2% | 1 | 0.2% | 1 |
| Trans-European | 3.4% | 2 | 3.9% | 2 |
| Turkey | 0.0% | 0 | 0.6% | 2 |
| Ukraine | 0.0% | 0 | 0.3% | 1 |
| UK | 11.2% | 12 | 9.0% | 8 |
| USA | 63.3% | 46 | 59.2% | 44 |
| Total | 100.0% | 112 | 100.0% | |

**Table 1.3.** *Share of main arms sales by country (source: SIPRI ranking of the world's leading defence companies)*

This increase in competition, linked to the development of other DTIBs, is also clear, particularly regarding naval military. For the past 10 years, we have witnessed the desire of countries such as Russia and China to develop their own maritime capacity, which has created new industrial competitors for companies already in place in this sector (Sheldon-Duplaix 2018).

Increased competition also comes from civil companies that have invested heavily in R&D. Among these companies, the GAFAMs that dominate the digital market play a major role. Indeed, R&D investments by digital giants are huge. Thus, in 2017, Amazon became the world leader in R&D expenditure, ahead of Alphabet, Google's parent company, and Intel, according to PwC. With Apple and Microsoft being ranked in the American Top 5, these companies spend, according to FactSet data, 76.2 billion dollars on R&D, while Facebook is in ninth place, thanks to a 32% increase in its R&D expenditure compared to 2016.

However, as Farhad Manjoo pointed out in 2017, not only are GAFAM R&D expenditures approximately the same as those of the American federal government (if military research is withdrawn) but more generally private companies in the technology sector "… do not only finance major projects, but also finance what will change the world"[5]. Their expenses are massively devoted to all the technologies that will be at the heart of our daily lives in the years to come and are based on artificial intelligence, such as autonomous cars, rockets, drones, personal assistants or virtual reality headsets.

In addition, as Chiva (2018) points out, the availability of disruptive technologies created by these private multinationals, the GAFAs and also the BATX (the Chinese Baidu, Alibaba, Tencent and Xiaomi) leads to the democratization of access to innovation, particularly in the digital field, which does not require any industrial production tool.

Artificial intelligence such as "big data" is freely accessible and allows any developer to create programs previously reserved for traditional defence stakeholders. Image recognition and big data mining are no longer reserved for defence manufacturers. Other fields are also undergoing a revolution, due to the emergence of techniques such as rapid prototyping, 3D printing or additive manufacturing. Finally, the most technologically advanced fields, hitherto considered as sovereign and reserved, are also developing, for example biotechnology. This is the case, for example, of new "DNA scissors" technologies, CRISPR-Cas9, which could be used by biohackers to manipulate genes.

The technological advances of these companies have increased the number of contracts with government, civil or defence agencies, such as the

5 "Google not the government is building the future", *New York Times*, 17 May 2017.

CIA or the United States Department of Defence, which have also created innovative partnerships with the civil innovation sector.

### 1.2.2.2. *Increasing complexity of research activities*

The increase in R&D activity noted above, as well as the increase in the internationalization of this activity, the number of stakeholders involved or the technologies mobilized, have led to the increasing complexity of research activities.

Indeed, technological complexity seems to increase over time, with each generation of technology building on the previous technological environment (Nelson and Winter, 1982; Howitt 1999; Aunger, 2010). Technology systems have therefore become increasingly complex over time due to the cumulative nature of knowledge and technologies. Technologies are also becoming more complex due to the increasing number of functions. Similarly, technologies have reached higher levels of complementarity requiring more multi-technology activities (Fai and Von Tunzelmann, 2001). The increase in technological levels in both the civilian and military fields has led to increasingly complex defence products, and has had a major impact on the developmental costs of these products (Hobday *et al.* 2005, Hartley and Sandler, 2012). The technological complexity induced by the mass application of new information and communication technologies (NICTs) to weapon systems is emblematic of the growing complexity of knowledge observed in the defence industry.

To take into account all of these changes, complexity indices have been created (Hausmann and Hidalgo, 2009). Broekel (2018) thus highlights these characteristics of technological complexity (increasing development over time, greater R&D efforts, more collaborative R&D, spatial concentration) that serve as benchmarks for the empirical evaluation of complexity. This evaluation uses European patent data from the years 1980 to 2013 and shows that the structural complexity measurement it proposes reflects the four typical characteristics as well as traditional measurements.

Moreover, as Gassman *et al.* (1999) note, there is a trend towards the internationalization of R&D, which although it became apparent in the 1970s, only became widespread in the late 1980s. This internationalization is mainly due to the desire to take advantage of technological advances developed in other countries, learning effects and the complementarity of skills. Thévenot (2007), in the case of France and based on a business relation survey[6], shows that complementarity with a R&D partner significantly favors the choice to internationalize R&D because it promotes learning. Access to new markets also plays a role in this, but to a lesser extent. Finally, cost reduction does not seem to be an important objective and there are very few relationships with stakeholders located in low-cost countries.

The growth in technological complexity and the internationalization of R&D activities have resulted in an increase in the number of stakeholders involved in innovation. This increase was also supported by the greater involvement of companies in the R&D activities highlighted above and by changes in the research method or production process. Indeed, defence companies have gradually reorganized their industrial connections since the early 1990s, with large firms repositioning their activities in the system integrator field (Hobday *et al.* 2005). This has led them to significantly increase the outsourcing of their R&D (Belin *et al.* 2018)[7] and to modify their subcontractor relationships, particularly in Aeronautics (Becue, Belin and Talbot, 2014). Faced with the increasing complexity of knowledge used to design and manufacture aircraft, aircraft manufacturers such as Airbus, Boeing, Bombardier or Embraer cannot master all systems. They have therefore gradually implemented strategies to refocus and outsource to strategic suppliers. The former are qualified as integration architects, the latter as pivotal firms for a major component. The pivot company is a coordinating firm that must manage a value chain (Fulconis and Paché, 2005) and exercise local leadership over a portion in the supply chain, in particular by project managing (Fabbe-Costes, 2005) or by controlling information flows through the management of different communication tools (Lorenzi and Baden-Fuller, 1995). The pivotal firm is positioned in a strategic segment of the supply chain and takes charge of research.

### 1.2.3. *Less dependence on defence financing*

At the same time, the R&D financing system has been significantly transformed in OECD countries since the 1980s. Financial markets have developed with the creation of markets or products dedicated to finance

6 Enquête sur les relations interentreprises (ERIE).

7 Companies receiving defence financing from external sources accounted for 13.4% of their R&D expenditure in 1987 compared with 22.5% in 2010, a share equivalent to that of non-defence companies (Belin *et al.* 2018).

innovation (derivative, alternative markets, etc.). New types of private forms of funding and institutions have emerged (such as venture capital or dedicated investment funds) and at the public level (creation or grouping of agencies, development of regional and European financing, etc.). These developments coincided with reforms of public support schemes for R&D activity.

Overall, the increase in public support for R&D has not kept up with the growth in R&D business spending despite the significant increase in the amount of such funding since 1981, particularly in the USA[8]. The proportion of government-funded business R&D expenditure has therefore declined since the mid-1980s, particularly in some European countries, such as France and the UK, as well as in the USA (Figure 1.6).

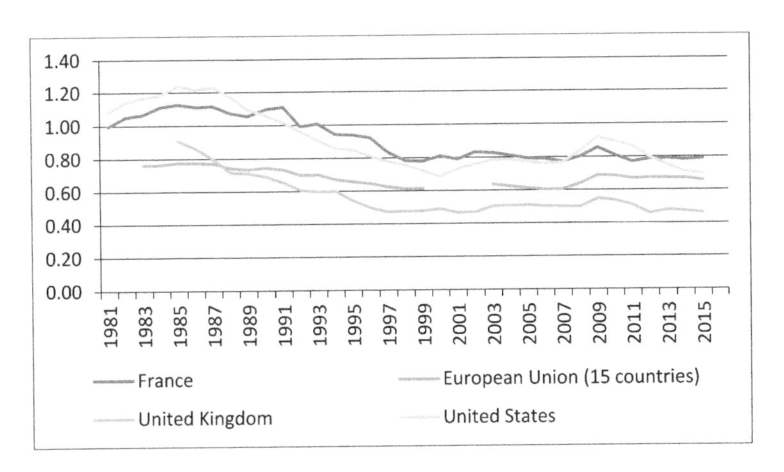


**Figure 1.6.** *State-financed GERD as a percentage of GDP*

In European countries, it is mainly direct aid that has been reduced, particularly defence aid, while in the USA, the downward trend reversed in 2001, following the attacks of 11 September, and allows defence financing to return to a level almost comparable to that at the beginning of the period as a proportion of Government Budget Appropriations or Outlays for R&D

8 The increase in this financing is 345.8% in the United States since 1981 compared to 263.5% for the European Union (15), 141.6% for France and 120.9% for the United Kingdom. Public funding for R&D therefore remains very high in the United States, where it represented $150,392 in 2016 compared with $113,883 in the European Union (15), $17,430 in France and $14,604 in the United Kingdom (current PPP).

(GBAORD): 51.9% in 2016 versus 54.6% in 1981 (Figure 1.7). The central role of the Department of Defence (DoD) in the US innovation system has therefore not been changed, with defence funding supporting R&D activities in industry and universities since the 1950s, as Mowery (2009) points out, still accounting for more than half of GBAORD in 2016.

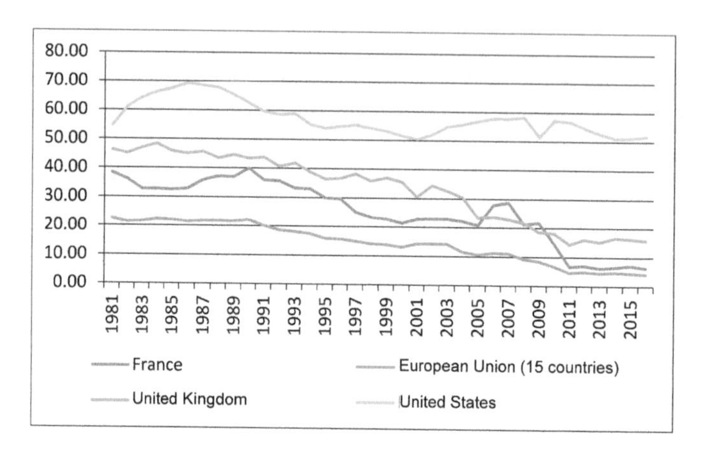


**Figure 1.7.** *Defence[9] R&D appropriations as a percentage of total GBAORD*

On the contrary, in European countries, the decline in direct aid, particularly defence aid, has been offset, at least partially, by the fairly significant increase in indirect aid in the form of tax incentives. In France, these changes in the support system for R&D by companies have been particularly noticeable. Public funding fell sharply from the early 1990s onwards, reflecting a gradual reduction in funding from the Ministry of the Armed Forces in a context marked by the end of the Cold War and budgetary restraints. The decline in defence financing, which was also interrupted in 2001, gave way to a period of stabilization and then of recovery, but resumed in 2009, following the effects of the 2008 financial crisis, which continued in Europe with a public debt crisis. At the same time, indirect *ex-post* aid increased from 2004, thanks to the development of a research tax credit, the *Crédit Impôt Recherche* (CIR), the amount of which exceeded that of direct civilian aid in 2005, and all direct aid, including defence, from 2008 stabilizing at approximately 0.26% of GDP since 2009.

9 "All state-funded defence R&D, including nuclear and space military R&D, but excluding civilian R&D funded by the Ministry of Defence (e.g. for meteorology)". Source OECD.

Thus, while overall, public R&D financing has returned to the level it was in the early 1990s since the end of the last decade, reaching 0.38% of GDP in 2011, the distribution of this financing has reversed (Giraud *et al.* 2014). As a proportion of business GERD, the total public funding amounted to 26% in 2011, with 8% direct funding and 18% CIR. Defence financing[10], which for a long time was twice that of civilian direct aid, at 0.17% of GDP in 1993, became a financing source only equivalent to 0.05% of GDP in 2011.

This structural reform of public support for R&D, which gave priority to indirect funding, was accompanied by the development of programs encouraging collaborative research: thematic projects of the ANR (National Research Agency), competitiveness clusters, etc.

These structural changes have brought the French model closer to those of other European countries that began to develop indirect financing earlier (Bodas-Freitas and von Tunzelmann, 2008) and particularly affected defence companies. Indeed, the defence financing that was at the heart of the system has become one of several involved in a policy that gives priority to indirect R&D financing through the CIR (Belin *et al.* 2018). The result is an environment that is more favorable to R&D investment for all companies regardless of their characteristics (in terms of size, age or sector) and gives them more autonomy in the selection of research projects and partners, which is justified in a context marked by innovation and rapid technological advances.

However, while the development of the CIR has quantitatively offset the reduction in direct public aid, particularly for the defence sector, it has not been sufficient to support the growth of enterprise R&D activity. In France, like in the European Union (except the UK) and the USA, companies have therefore had to finance this growth by investing more in their R&D activity (Figure 1.8).

10 The Business GERD amounted to €28.8 billion in 2011 for total public financing of €7.4 billion, including €2.3 billion in direct financing and CIR 5.2 billion.

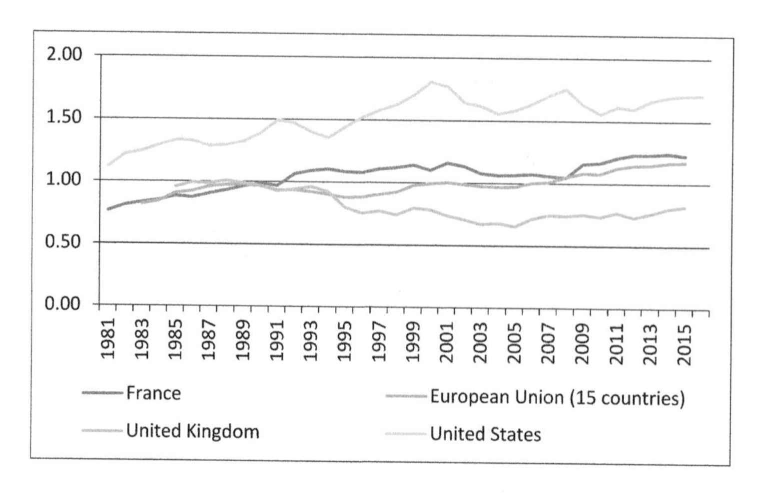


**Figure 1.8.** *Business-financed GERD as a percentage of GDP*

On the contrary, qualitatively, the same types of projects are not funded. *Ex ante* direct aid is more targeted and while it has the disadvantage of assuming that public authorities are able to identify relevant sectors and projects, it is more suitable for projects with particularly high externalities or risks. Ministry of the Armed Forces aid often finances major contractors involved in this type of project, and basically covers defence R&D, while the development of non-targeted CIR makes it possible to increase the number and diversity of beneficial companies (particularly in terms of size and sector)[11].

All of these developments have made it possible to finance the growth of companies' R&D activity while making them less dependent on direct public funding, particularly for defence. However, defence companies and the Ministry of the Armed Forces have had to adapt to this new environment, as in other countries.

11 Between the first reform to develop the CIR in 2004 and 2011, the number of companies benefiting from the CIR more than doubled. And these beneficiary companies are mainly SMEs covering a wide range of sectors. Thus, of the 15,000 who benefited from it in 2011, only 12% were large companies, and they received about a quarter of this indirect funding. On the contrary, large companies received almost all public defence financing (91%) in the same year (Giraud *et al.* 2014).

## 1.3. Opening of the defence sector

All actors in the defence sector, the Ministry, agencies and companies, have opened up to the civil sector. This openness makes it possible to encourage the development of networked innovations with the decompartmentalization of the civil and military sectors. It is manifested through the development of duality, cooperation and the emergence of new institutions to strengthen links with civil research.

### 1.3.1. *Development of duality*

The concept of duality, which plays a major role in defence, refers to a multitude of different realities (te Kulve *et al.* 2003). Civil/defence duality has two dimensions: on the one hand, the existence of two elements of a different nature and, on the other hand, the existence of complementarity. This complementarity can be scientific, industrial, commercial, legal, social and financial, etc. Duality can be a characteristic of products, knowledge, skills or technologies, as well as of a company or even of the national information system (NIS) as a whole. However, it is difficult to measure empirically.

The evolution of the duality of defence companies can be analyzed based on the evolution of the total defence revenue/total revenue ratio. However, this analysis is practically impossible to carry out due to the lack of available data, in particular due to the significant concentration movement observed in the defence sector since the end of the Cold War. This restructuring of the defence industry took place in the 1990s in the USA (Harper, 1999) and was slower in Europe where it is still ongoing. In France, we can mention in particular the privatization of Thomson-CSF in 1998, the creation of Thales and EADS in 2000, MBDA in 2001, SAFRAN in 2005, Nexter in 2006, etc. Moreover, we do not have any data for the ratio of defence turnover to total turnover.

The evolution of duality has also differed from one field to another. It is very high in aeronautics or electronics and remains very low in naval, land or missile applications. It is also different depending on the company's place in the production chain (principal or subcontractor) but here again, figures are scarce.

On the contrary, the change in the share of defence activities does not give any indication of existing collaborations at the scientific level, where complementarity may be greatest. According to Guichard (2005), the relationship between the civil and defence sectors in research and innovation can be broken down into three main phases:

– the operating pattern of the golden age of the 1960s with a strong programmatic logic, a powerful client state and a dynamic driving force in the national economy. Much more R&D activity and funding in the defence sector than in the civilian sphere, and therefore a strong dependence of the civilian sector on the military sector;

– an isolation of defence research and an "off-the-shelf" acquisition that makes it possible to benefit from innovations and civilian products. However, this strategy requires an intelligent buyer with the skills to choose the right product or technology. There is a strong dependence of the military here on civilians;

– a networked innovation with a decompartmentalization of the civil and military sectors that makes it possible to achieve both strategic objectives at the military level, as well as objectives in terms of innovation and growth.

In France, the concept of duality has been very present since the late 1990s (Mérindol and Versailles 2014), even though one of the characteristics of the French defence technological system was the existence of discontinuity between the user, industry and science. France was rather in phase 2, with a certain isolation of defence research. It currently appears to be moving into phase 3.

On the contrary, changes in the American system have been faster and the civil–military decompartmentalization has been greater. In the USA, the concept of duality appeared after World War II in speeches on the export of weapons and military technologies, regarding the creation of the Coordinating Committee for Multilateral Export Controls, CoCom, in 1949 (Reppy 1999). At first, this concept had a rather negative connotation and duality was seen as a source of difficulties and problems. In such cases, it was necessary to prevent the spread of technologies and weapons that could give States a strategic advantage. At the end of the Cold War, the doctrine changed, technological duality was encouraged and sought. It would make it possible to maintain a high technological level in terms of defence despite the decline in budgets (between 1989 and 1999 the acquisition budget decreased by 52%, Depeyre 2013), while increasing the country's competitiveness through a more efficient allocation of public R&D funding. The American government thus promoted the decompartmentalization of civil–military relationships while maintaining defence R&D at the center of the American NIS (Guichard 2004). It sought, on the one hand, to increase cooperation between the government, industry and universities and, on the other hand, to promote the spread of technologies from defence to civil and from civil to defence sectors. Financing plays an important role in the management of duality. Feldman (1999) and Guichard (2004) thus highlight the role of the Pentagon in the emergence of biotechnology in the USA, with significant funding in the 1980s and 1990s to companies in the sectors that did not develop specific military applications.

The American military technological system has therefore always been characterized by strong relationships between state–science–industry. One of the differences between the USA and France is the greater involvement of universities in defence-related research processes (Kenney and Mowery 2014)[12]. Mérindol (2004) thus points out that "in Europe, the knowledge networks in which defence operates have been marked by the restructuring of stakeholders and coordination methods, while in the USA, the interaction between the State, science and industry has never been called into question".

### 1.3.2. *More cooperation*

The development of duality within the system requires more cooperation so that civil/defence complementarity can develop. As Mérindol (2004) points out, the complexity of the innovation process and its collective and interactive nature make scientific and technological networks and intermediation a decisive element of research and innovation policies in their civil and military dimensions.

12 As Mérindol (2004) points out, more than half of the fundamental research funded by the DoD is entrusted to universities. The mobility of staff between the public and private sectors is also very important.

The development of civil innovation has made it necessary to strengthen cooperation between defence and civil stakeholders. At the same time, the intensification of R&D efforts has made innovation processes more complex by mobilizing more knowledge and stakeholders. It has also accelerated innovation processes. Research and innovation prices have therefore increased in defence, forcing governments and companies to develop cooperation in order to reduce the costs of developing weapons programs (Hartley 2007).

As a result, defence companies and ministries of defence have had to open up and cooperate more with civilians. Cooperation between companies and between States has also developed.

As Malizard (2015) points out, all defence companies are seeking to set up such cooperation. At the state level, we can mention the Sicral 2 satellite project for the Italian Ministry of Defence and the Direction Générale de l'Armement (DGA)[13], which was developed by Thales Alenia Space, via a cooperative program led by the Italian and French Defence Ministries under a bilateral agreement. More generally, Serfati (2014) indicates that 30% of French weapons programs are carried out in cooperation.

French defence companies have also developed their cooperation with other companies or research centers. Thus, in addition to their own research centers (Thales Research and Technology or Safran Tech), defence groups are partners with many laboratories and research chairs. Safran and Thales are members of IRT 5 (Institut de Recherche Technologique) Antoine de Saint-Exupéry, which specializes in aeronautics, space and embedded systems issues. Innovative methods of innovation management are also being implemented. For example, MBDA is developing an innovation college to improve the efficiency of the innovation process and the IDEA collaboration program available on the company's intranet, which aims to increase interactions and ideas between researchers and employees. Thales applies the Design Thinking model, whose aim is to better define the problems posed by the innovation process rather than to focus on solutions because a poorly defined problem leads to erroneous solutions. MBDA practices improve innovation by organizing meetings with innovative SMEs,

13 French Defence Procurement Agency.

university laboratories and other industrialists. For example, MBDA is piloting the British MCM ITP (Materials and Components for Missiles, Innovation Technology Partnership) project with the French and British Ministries of Defence for R&T work on TRL 1 to 3 missiles, in association, in particular, with Safran and Thales. At Safran, a structured and systematic Group-wide innovation approach enables large-scale projects to emerge, such as the "electric taxiing system" for commercial aircraft, currently being developed in cooperation with Honeywell. Finally, an Economic Interest Grouping (EIG) has been set up between Thales, CEA and Alcatel-Lucent to develop technologies for semiconductor components.

### 1.3.3. *New institutions and stronger links with academic research*

As a result of these developments, the defence sector has had to expand to include the civilian domain. It no longer controls the financing and execution of R&D and research activities have become more complex. It no longer controls most of the information concerning technological or competitive developments. It can no longer be vector through which information can be transmitted, but must seek this information and collaborate with all public, private, defence and civilian stakeholders.

As noted above, the American defence system has always sought to maintain strong state–science–industry relationships by adapting to technological change. Thus, in 1999, the CIA created the In-Q-Tel venture capital investment fund, which is responsible for financing start-ups best able to develop technologies useful for intelligence. Among its most emblematic investments is Google Earth. In 2015, the Pentagon created a laboratory called the Defence Innovation Unit Experimental (DIUx) with the objective of capturing innovation from the civilian world at the R&D stage and transferring it to defence (Chiva 2018). Located at the heart of American start-ups and high-tech industries, and not in Washington, within the American power and defence ecosystem, the DIUx finances projects identified with the various DoD entities. The objective is to share the costs, as the company has already invested in its R&D, which is no longer State responsibility. DIUx only finances pilot projects and in a different way from traditional equity investments (non-dilutive financing). If the project is successful, the costs of acquiring the technology are covered by the DoD entity that wishes to deploy it (for every dollar spent by DIUx, the entity concerned pays, on average, three dollars).

This example shows that with a relatively modest budget ($29.6 million in 2018), even though it can be supplemented by private venture capital, it is possible to transfer civil innovations to defence, provided that new operating modes are developed. Indeed, it is necessary to adapt to the culture of start-ups, in particular to accept the acceleration of pace, risk-taking, the possibilities of failure, and to rethink contractual mechanisms allowing the purchase of innovations. According to Doherty (2015), these transformations correspond to a "collective disruption": the agency acts as an intermediary between the US Ministry of Defence, which plays an incentive role, on the one hand, and SMEs and start-ups, which are the innovators, on the other hand.

In France, the creation, in 2018, of the Defence Innovation Agency reflects the same desire to maintain technological sovereignty. The appointment of Emmanuel Chiva, mentioned above and familiar with the American research system, as Director reflects the desire to establish new relationships with the civilian world. The Defence Innovation Agency, which is directly attached to the Delegate General of the DGA, is intended to be the central player in the Ministry of the Armed Forces's new innovation strategy, in conjunction with Europe and visible at the international level, as Florence Parly, Minister of the Armed Forces, pointed out in 2018[14]. This agency therefore reflects a greater openness of defence towards civilians, as well as a desire to better coordinate the stakeholders.

To facilitate this development, to increase links and knowledge of the latest technologies or to preserve strategic skills, the DefInvest fund was also created in 2018. By financing innovative French start-ups and SMEs that have the potential to develop breakthrough technologies that hold great promise in terms of military applications, it aims to increase the efficiency of the system by making it possible to achieve both strategic objectives at the military level and objectives in terms of innovation and growth. The fund will benefit from the complementary expertise of Bpifrance, a subsidiary of Caisse des Dépôts (CDC) and the French State, and the DGA. This fund reflects the openness of the defence sector in two ways. First, the Ministry of the Armed Forces does not create this fund alone but in association with Bpifrance. Bpifrance operates in the civil sector by supporting companies from the initial public offering to stock exchange listing, credit, guarantee

14 Speech at the MEDEF Summer School, August 28, 2018.

and equity financing. Then, this fund opens up to innovative French start-ups and SMEs, whether they are defence or civil[15]. From an initial amount of 50 million euros, the fund should grow in the coming years.

The French Ministry of the Armed Forces is also modifying its relations with the Universities and the CNRS in order to strengthen their links with the defence sector. The latest French White Paper on Defence and National Security (LBDSN) stressed in 2013 that "the State's forward-looking approach must be based on independent, multidisciplinary and original strategic thinking integrating university research". The Ministry of the Armed Forces, the CNRS and the Conference of University Presidents therefore signed an agreement in January 2017 to expand research on defence and strategy issues. The Ministry has also implemented the Higher Education Pact. This mechanism aims to help regenerate the pool of university research in the fields of defence and security. These initiatives are once again a symbol of a greater openness of defence towards civil society and an increase in cooperation between both. Until now, French universities have not been very involved in defence research and it has been carried out in research centers or schools under supervision.

The inclusion of civil society and the increase in cooperation are also reflected in the structures and policies implemented at the European or national level. Thus, in July 2018, the European Parliament approved the establishment of a European Defence Fund in 2019. This fund should make it possible to develop the military capabilities of the Member States and the strategic independence of the European Union. The future European Defence Research Programme should provide structural funding for defence research of €20 billion for the period 2021–2027, including €13 billion for the European Defence Fund (€8.9 billion for development and €4.1 billion for research). The objective of this fund is to finance projects set up in cooperation with the presence of at least three companies from at least three Member States in each project. Moreover, 10% of the budget should be allocated to SMEs. These budgets will also be reserved for European companies: based in Europe and owned by European capital.

15 Target companies are "companies with products, technologies or know-how essential to the performance of French defence systems or start-ups, SMEs or ITEs with an innovative project whose outlets are potentially disruptive for the French defence systems of tomorrow".

## 1.4. Conclusion

From the 1950s to the end of the 1980s, the State, and more particularly the Ministry of Defence, was largely in control of the national innovation systems of the main developed countries. A significant share of public R&D funding was distributed by defence and very few other sources of funding (public or private) for research and innovation existed. A significant proportion of the research was carried out by the state. Most defence companies were also public. This structure made it easier for the State to manage the domain. Defence was in a dominant position, knowledge about technologies was increasing and defence technologies were spreading to the civil sector. Innovation processes were also simpler, involving fewer technologies and stakeholders and their rates were slower.

Since the late 1980s, we have witnessed significant changes in NISs, particularly in France. Innovation, R&D and civil funding have increased significantly. The French Ministry of the Armed Forces is no longer a dominant player in the financing of enterprise R&D, it only grants a very small proportion of the public funding to them (6% in 2016). At the same time, companies and universities have increased their role in French R&D and the share carried out by the State has fallen sharply. Civil R&D has grown strongly from 1.5% of GDP in 1981 to 2.2% in 2014. The intensification of research efforts has made innovation processes more complex by mobilizing more knowledge and stakeholders. It has also accelerated innovation processes.

These changes in the French innovation system have called for a restructuring of the defence sector and greater cooperation with civil society. Defence companies are still as much involved in research and innovation as ever, but the defence sector has gradually had to open up to other ways of funding and other stakeholders (civil companies, universities, European partners). We seem to be moving into a phase characterized by a more significant decompartmentalization of civil and military affairs, which allows the defence sector to share its historical advantages with the civilian sector: to help coordinate stakeholders, to research and analyze technological information and to finance disruptive technologies.

## 1.5. References

[AUN 10] AUNGER, R., "Types of technology", *Technological Forecasting and Social Change*, vol. 77, no. 5, pp. 762–782, 2010.

[BEC 14] BECUE, M., BELIN, J., TALBOT, D., "Relational rent and underperformance of hub firms in the aeronautics value chain", *Management*, vol. 17, no. 2, pp. 110–135, 2014.

[BEL 18] BELIN, J., GUILLE, M., LAZARIC, N., MÉRINDOL, V., "Defence firms adapting to major changes in the French R&D funding system", *Defence and Peace Economics*, 2018.

[BRO 18] BROECKEL, T., "Measuring technological complexity – Current approaches and a new measure of structural complexity", *arXiv.org Working Paper arXiv:1708.07357*, 2018.

[CHI 18] CHIVA, E., "Defence innovation unit experimental", *Défense et Industries*, no. 11, FRS, 2018.

[COL 15] COLATAT, P., "An organizational perspective to funding science: Collaborator novelty at DARPA", *Research Policy*, vol. 44, no. 4, pp. 874–887, May 2015.

[DEP 13] DEPEYRE, C., "Boeing boeing: la dualité civil-militaire source d'un rebond stratégique dans l'ère post-Guerre Froide", *Entreprises & Histoire*, no. 73, pp. 58–74, 2013.

[DOC 15] DOCHERTY, M., *Collective Disruption: How Corporations and Startups Can Co-Create Transformative New Businesses*, Polarity Press, Boca Raton, 2015.

[FAB 05] FABBE-COSTES, N., "La gestion dynamique des supply chains des entreprises virtuelles", *Revue Française de Gestion*, vol. 31, no. 156, pp. 151–166, 2005.

[FAI 01] FAI, F., VON TUNZELMANN, N., "Industry-specific competencies and converging technological systems: Evidence from patents", *Structural Change and Economic Dynamics*, vol. 12, no. 2, pp. 141–170, 2001.

[FUC 10] FUCHS, E., "Rethinking the role of the state in technology development: DARPA and the case for embedded network governance DARPA", *Research Policy*, vol. 39, no. 9, pp. 1133–1147, November 2010.

[FUL 05] FULCONIS, F., PACHE, G., "Piloter des entreprises virtuelles: un rôle nouveau pour les prestataires de services logistiques", *Revue Française de Gestion*, vol. 3, no. 156, pp. 167–186, 2005.

[GAS 99] GASSMAN, O., VON ZEDTWITZ, M., "New concepts and trends in international R&D organization", *Research Policy*, pp. 231–250, 1999.

[GIR 14] GIRAUD, L., MIOTTI, L., QUEMENER, J. *et al.*, Développement et impact du crédit impôt recherche: 1983–2011, Rapport MESR, 2014.

[GUI 05] GUICHARD, R., "Suggested repositionning of Defence R&D within the French system of innovation", *Technovation*, vol. 25, no. 3, pp. 195–201, 2005.

[HAR 99] HARPER, J.K., "Corporate governance and performance during consolidation of the United States and European defence industries", *Journal of Management and Governance*, vol. 2, pp. 335–355, 1999.

[HAR 07] HARTLEY, K., "The arms industry, procurement and industrial policies", in SANDLER, T, HARTLEY, K. (eds), *Handbook of Defence Economics*, vol. 2, North-Holland, 2007.

[HID 09] HIDALGO, C.A., HAUSMANN, R., "The building blocks of economic complexity", *Proceedings of the National Academy of Sciences of the United States of America*, vol. 106, no. 26, pp. 10570–10575, 2009.

[HIR 12] HIRSCHEY, M., SKIBA, H. WINTOKI, B., "The size, concentration and evolution or corporate R&D spending in US firms from 1976 to 2010: Evidence and implications", *Journal of Corporate Finance*, vol. 18, no. 3, pp. 496–518, 2012.

[HOB 05] HOBDAY, M., DAVIES, A, PRENCIPE, A., "Systems integration: A core capability of the modern corporation", *Industrial and Corporate Change*, vol. 14, no. 6, pp. 1109–1143, 2005.

[HOW 99] HOWITT, P., "Steady endogenous growth with population and R & D inputs growing", *Journal of Political Economy*, vol. 107, no. 4, pp. 715–730, 1999.

[KUL 03] TE KULVE, H., WIM, S.A., "Civilian–military co-operation strategies in developing new technologies", *Research Policy*, vol. 32, no. 6, pp. 955–970, June 2003.

[LAZ 11] LAZARIC, N., MERINDOL, V., ROCHHIA, S., "Changes in the French defence innovation system: New roles and capabilities for the government agency for defence", *Industry and Innovation*, vol. 18, no. 5, pp. 509–530, July 2011.

[LOR 95] LORENZI, G., BADEN-FULLER, C., "Creating a strategic center to manage web of partners", *Californian Management Review*, vol. 37, no. 3, pp. 147–163, 1995.

[LUN 92] LUNDVALL, B.A., "Introduction" in *National Systems of Innovation*, LUNDVALL B.A. (ed.), Pinter, London, 1992.

[MAL 15] MALIZARD, J., "L'innovation comme facteur de croissance, l'exemple de grands groupes industriels de défense français", *Revue Défense Nationale*, no. 780, May 2015.

[MER 04] MERINDOL, V., "R&D de défense et coordination civile-militaire", in UZUNIDIS, D. (ed.), *L'innovation et l'économie contemporaine*, De Boeck, pp. 85–113, 2004.

[MER 14] MERINDOL, V., VERSAILLES, D., "La dualité dans les entreprises de Défense Française [The Management of Dual-use Technologies in French Defence Firms]," Observatoire Economie de la Défense, Ministère de la Défense. Available at : http://www.defence.gouv.fr/dgris/recherche-et-prospective/etudes-prospectives-et-strategiques/eps-2013-2014, 2014.

[MOW 12] MOWERY, D.C., "Defence-related R&D as a model for "Grand Challenges" technology policies", *Research Policy*, vol. 41, no. 10, pp. 1703–1715, 2012.

[NEL 82] NELSON, R., WINTER, S.G., "An evolutionary theory of economic change", *Cambridge (Mass.)*, Belknap Press/Harvard University Press, 1982.

[NIO 93] NIOSI, J., BELLON, B., SAVIOTTI, P.P. *et al.*, "National systems of innovation: In search of a workable concept," *Technology in Society*, vol. 15, no. 2, pp. 207–227,1993.

[OCD 18] OCDE, "Principaux indicateurs de la science et de la technologie", *Statistiques de l'OCDE de la science et technologie et de la R&D* (database). Available at: https://doi-org.docelec.u-bordeaux.fr/10.1787/data-00182-fr (data accessed 2 September 2018).

[PAP 14] PAPAGEORGIADIS, N., CROSS, A.R., ALEXIOU, C., "International patent systems strength 1998–2011", *Journal of World Business*, vol. 49, no. 4, pp. 586–597, 2014.

[PAR 08] PARK, G.W., "International patent protection: 1960–2005", *Research Policy*, vol. 37, pp. 761–766, 2008.

[SAN 07] SANDLER, T., HARTLEY, K. (eds), *Handbook of Defence Economics*, vol. 2, North-Holland, 2007.

[SER 14] SERFATI, C., *L'industrie française de défense*, La Documentation française, 2014.

[SHE 18] SHELDON-DUPLAIX, A., "Implications des ambitions maritimes des États puissances et des innovations navales", *Recherches & Documents de la FRS*, 04 February, 2018.

[THE 07] THEVENOT, C., "Internationalisation des réseaux de R&D: une approche par les relations d'entreprises", *Economie et Statistique*, vol. 405–406, pp. 141–162, 2007.

# 2

# Evolution of the Aerospace and Defence Innovation Model: Intensifying Science and Technology Relationships

ABSTRACT. This chapter looks more closely at the evolution of innovation model in aerospace and defence (A&D) for a long period (1945–2015). Our main goal is to show the emergence and development of the knowledge-based economy through an analysis of innovation models. To do that, we develop an empirical analysis of patents in technological fields of A&D. This study points to a constant growth of non-patent literature in patent citations since 1980. In order to deepen this analysis, we study the correlation between non-patent literature citation and the patent quality in A&D technological fields and we find a clear development in innovative practices, which include more and more diversified knowledge, particularly scientific knowledge.

## 2.1. Introduction

On June 7, 2017, the European Commission presented its vision of "European Defence" centered around scientific and industrial research. This project is a first step towards developing a common defence innovation policy for the Member States of the European Union through an annual budget estimated at 500 million euros per year. This progress in European cooperation in terms of research should be accompanied by results in terms of innovation. However, the defence innovation model is often approached by linear models, which are now being challenged by the increasing production and sharing of knowledge in society brought about by the revolution of new information and communication technologies. A new

Chapter written by Cécile FAUCONNET.

economic model is emerging: that of knowledge-based economy which emphasizes in particular the place of scientific research in the innovation process.

In this chapter, we examine the evolution of the innovation model of the aerospace and defence (A&D) sector in light of this vision of the economy. A&D is no exception to the evolution of innovation models and more specifically in terms of the use of scientific knowledge. First, we analyze the evolution of the contribution of scientific knowledge to technological innovations in A&D between 1945 and 2015. Second, we look at the influence of this evolution on the quality of inventions in this sector.

To realize this study, we use bibliometric tools. The intensity of use of scientific knowledge in innovation is introduced by Narin's research [NAR 85, ALB 91, NAR 97]. He studies the intensification of the relationship between public scientific research and American technological production through two vectors: on the one hand, through the publication of patents by academic researchers and, on the other hand, through total *non-patent literature* (NPL) present in patents over two periods (1987–1988 and 1993–1994) [NAR 97]. Their paper shows a 30% increase in NPL citations in patents between these two periods, i.e. a significant increase in the use of scientific knowledge for the production of innovation. In addition, authors such as [MOW 01] or [SAM 03] point out that patents citing fundamental knowledge are more original and sensitive to influence from other technological categories. According to [TIJ 04], the NPL analysis represents the best approximation of science and technology (S&T) interactions and human creativity, allowing industrial innovation. We integrate our analysis of the links between S&T into this tradition by focusing on the production of A&D. We identify patents filed in the technological fields of A&D through the classification of the Observatoire des Sciences et des Techniques [OST 10][1]. These fields include five technological classes corresponding to offensive and defensive installations on ships, astronautics, explosives, weapons and ammunition (see Appendix 1).

Thus, in the first part of this chapter, we review the literature to establish the theoretical framework of our reflection and the particularity of our field of application, A&D. The following sections aim to test our theoretical questioning, which can be summarized as follows: does the emergence of the

1 See Appendix 1.

knowledge-based economy resulting in increased collaboration between producers of scientific and industrial knowledge and leading to increased production of technological innovation, both in quantity and quality? The second part of our work is devoted to the patent data used for this study and a description of the method used. Finally, the final section will discuss the evolution of the contributions of scientific knowledge to A&D innovation as well as a joint analysis of the quality of patents and the contributions of scientific knowledge to them in order to study the evolution of the A&D innovation model.

## 2.2. Reflection framework

### 2.2.1. *Defence innovation*

Technological innovation is defined in the Oslo Manual [OEC 05, p. 37] as "technologically new products and processes as well as significant technological improvements in products and processes that have been achieved". For defence activities, [SEM 15] defines innovation as the action or process of introducing new ideas that is essential to accomplish defence missions. Such capabilities allow defence missions to be carried out effectively, without being compromised by a potential adversary. The need to surpass potential enemies means that the armed forces must constantly improve their tactics and equipment. Thus, the defence industry has its own specificities because it carries within it strategic and sovereignty issues [DUP 13].

Today, defence innovation involves the development of solutions with a high degree of complexity, integration and economic value [DIT 06]. These intensive engineering solutions integrate a large number of technologies and require a set of complementary skills. A salient example is the fighter aircraft because it requires combinations of knowledge that differ drastically from civil sector innovation. Indeed, it requires skills in terms of engines (jet engine, rocket engine), electronics (radar, navigation system, electronic assistant, etc.) as well as armament-related technologies (air-to-air missile, air-to-ground, laser guided bomb, etc.).

Innovation uses two mechanisms to achieve its objectives [MAR 91]. The first is to create variety by exploring the different paths of knowledge creation (research, invention, development, design, etc.). The second focuses on the choice and exploitation of different alternatives. In the civil sector, creating variety is carried out by entrepreneurs whose objective is to make a profit following the introduction of a new product on the market. Innovation follows a "classic" process where R&D teams develop the design and the market provides a quick response to this new product. In the field of defence, incentives and profits follow a different path. First, equipment life cycles are long and innovation is "generational" [BEL 17]. Indeed, the paradigm of innovation in defence is techno-centric: the principle is the development of capabilities by successive generations in order to avoid any strategic surprises (examples being the Mirage 2000 and the Rafale). To take the example of combat aircraft, the launch of the "Rafale" project to debut in the 1970s, leading to the first commissioning for the French Navy in 2002 and the French Air Force in 2006. Second, the context in which the system will be used cannot be accurately anticipated. By means of illustration, the Tiger attack helicopters used by the French Army are not adapted to the natural conditions of certain Malian areas. Indeed, even though this helicopter is intended for use in the desert, the fineness of the sand in Mali leads to overwear of the engines due to filtration problems[2].

Military equipment has three main characteristics that strongly influence innovation: (i) military competition between states; (ii) differentiated production; and (iii) technological complexity [SEM 15]. First, competition of military equipment is based on several criteria: accuracy, reliability, maneuverability, protection, quality and interoperability. Such requirements require the complex development of the supply side, and a combination of scientific and technological research. Second, differentiated production is defined by the fact that States and their armed forces have different preferences in terms of war materials, due to geographical differences, types of potential conflicts, etc. When national preferences differ significantly, grouped purchases seem to be compromised. Finally, the complexity of military equipment is increasing with each new generation. It is due to the large number of subsystems, components and interfaces that make up the

2 Source: article from *Le Point International,* June 2, 2016. http://www.lepoint.fr/monde/mali-les-helicopteres-francais-de-l-operation-barkhane-les-pales-du-desert-02-06-2015-1932966_24.php#.

weapon systems. Indeed, while an automobile contains about 4,000 parts, a guidance system for an intercontinental missile contains more than 19,000 parts [KEL 1995] and a jet engine can exceed 22,000 parts [HOB 98].

The defence sector therefore has a special relationship with technological innovation. The arms race of the 20th Century led to dynamic innovation in society as a whole. An example of these innovations is GPS location, a project initiated in 1960 by the US military under the leadership of US President Richard Nixon. However, the end of the Cold War combined with the decline of military budget [FAU 16] led defence technology companies to review their innovation model and direct their production towards more duality [DEP 14] and thus adapt to "civilian" innovation models. The defence innovation model is often approached by linear models where innovation is perceived as a quasi-external process to the development of weapons programs [GAN 92] and where the contribution of scientific knowledge is distant and long-term. However, the evolution of the economy and society is challenging this linear vision of innovation.

### 2.2.2. *Knowledge-based innovation*

From the 1980s onwards, the increased use of intangible capital combined with the revolution of information and communication technologies led to the development of a knowledge-based economy [FOR 98]. The linear innovation models previously used no longer seem to be able to capture the reality of exchanges. Innovation no longer emerges in clusters or by surprise, but has become the norm. It is necessary for companies and becomes, in some sectors, the core of competition [LEN 02]. This perpetual technological change is reflected in a multilateral, multidisciplinary and changing combination of scientific and technological research and methods of invention and organization. In 1996, in Paris, the OECD [OEC 96, p. 3] stated that developed economies "are increasingly relying on knowledge and information". Knowledge is now recognized as a driver of productivity and economic growth, shedding new light on the role of information, technology and learning in economic performance. In particular, the place of scientific research in the innovation process is becoming increasingly prominent. [LEY 95] shows that globalization combined with the emergence of the knowledge-based economy is challenging the place of scientific research in society and more particularly its contribution to innovation. Science is no longer in its "ivory tower" with a long-term or even very long-term contribution to industrial innovation. It highlights a model where the transmission of knowledge between public and industrial research is dynamic and where scientific contributions can be short-term.

This evolution of the economy and the innovation process is theorized by [ETZ 00]. They are inspired by the evolutionary theory of the firm [NEL 85]. Instead of focusing on the co-evolution of firms and technology, knowing the architecture of knowledge, they add a historical perspective and thus endogenize the architecture of society. This model, called the *triple helix model* (THM), reflects the expanding role of knowledge in the relationship between political and economic institutions in society. Historically, systematic interactions between the market and science date back to the mid-19th Century and are best described by the concept of the "techno-scientific revolution" [BRA 74]. It is defined as the transformation of science itself into capital [NOB 79]. For the historian [BRE 02], collaboration between science and defence began in France in 1793 with the creation of the first public research unit, responsible for "researching and testing new means of defence". Thus, A&D, with its high-tech characteristics, complex systems and strong relationship between public decision-makers and the industry, is particularly relevant here. The THM, taking into account the relationship between science and the market, as well as between public and private companies, builds an analysis of interactions between public scientific research and industry. The various interactions between industry, government and universities generate new structures (i.e. public–private research centers, innovation poles) and create new mechanisms for integrating knowledge between the different stakeholders.

The starting point of this analysis is that the economy is evolving towards an increasingly intensive use and integration of knowledge through the dissemination of knowledge and collaboration of the various stakeholders in the growing society. In this chapter, we examine whether this observation also applies to the A&D industry despite its particularities. Our argument can be summarized as follows.

Does the emergence of the knowledge-based economy resulting in an intensification and increased codification of collaborations between producers of scientific and industrial knowledge, leading to a greater production of technological innovation, both in quantity and quality?

To do this, we need to study the links and their evolution between scientific and technological knowledge.

## 2.3. Methodology

To this end, we use patent data by combining two analytical tools: on the one hand, bibliometrics, which makes it possible to study the evolution of scientific literature citations, and on the other hand, patent quality indicators, to be correlated with the intensive use of scientific knowledge.

### 2.3.1. *Bibliometric approach*

The emergence of the knowledge-based economy creates the need to describe the distinctions between science and technology (S&T), between universities and industry. However, assessing the impact of research on innovative production is one of the most difficult fields of scientific analysis to investigate empirically [COO 98]; [SAL 01]; [COH 02]. Although there is diversity in approaches to the interface between S&T, a consensus is emerging on the empirical analytical method to be used: bibliometrics. This is a quantitative study of aspects of the scientific research process as a communication system [MIN 15], capable of providing measurements of the link between S&T at the *micro, meso and macro* scales. To a *micro* level, it captures the relationship between a scientific publication and a patent, which makes it possible to identify the proximity between different research objects. At the sectoral (or *meso*) level, it captures knowledge transfers and potential induced effects (description of the knowledge bases associated with a particular technology or industry). Finally, the *macro* level makes it possible to build concordance tables between scientific classifications and technological nomenclatures, to investigate dependencies and to develop public policy instruments.

Bibliometrics is a multifaceted tool designed to identify, quantify and analyze knowledge production and its qualities. Mostly used for the study of scientific literature, this method also allows the analysis of knowledge transmission channels between scientific and technological research. The knowledge flows enabling innovation are diverse: science to science, technology to technology, technology to science and science to technology. Since the 1960s, Narin has devoted his research to the analysis of knowledge production and dissemination, introducing a co-citation analysis, within scientific research (via scientific publications), technological R&D (via patents) as well as the flows between these two. The increase in scientific knowledge can arise from the knowledge contained in the scientific literature and that of technology by the knowledge disseminated in patents.

The method of analyzing citations between different knowledge support is the most widely used and it is this that we retain in our study. Presented by [NAR 82], it involves studying references to scientific articles (NPL) in patents and vice versa. According to [TIJ 04], the NPL analysis represents the best approximation of S&T interactions and human creativity allowing industrial innovation. "If the scientific literature reflects scientific activities, a multidisciplinary catalogue of citations can provide an interesting overview of these activities. This vision can highlight both the structure of science and the development of scientific processes" [GAR 79 p. 62]. This type of analysis can be extended to patents. Since the early 1980s, [CAR 81] has shown the value of NPL analysis in studying innovation processes. They highlight the dependence of technological change on scientific research. Other authors, such as [LET 07], are interested in specific technological areas. They show a positive relationship between the scientific intensity of patents and technological productivity in the biotechnology sector. This approach also allows analysis in terms of knowledge flows over a given geographical area, [VER 03] the geographical distribution of NPLs in the biotechnology and information technology sector and highlights national citations as well as the significant contribution of knowledge produced in the USA and Europe to technological innovation worldwide. From another perspective, [LOO 03] explain innovation performance by the intensity of use of scientific references in patents. In order to refine this type of analysis, we can identify three categories of citations: i) self-citation [NAR 97], ii) national citations and iii) international citations [HIC 01, NAR 97] thus highlights the bias of this analysis by highlighting the problem of *self-citation*, language barriers leading researchers to prefer national references (which are not necessarily the most relevant), and knowledge problems of patent applicants and authors of scientific articles on the production of research types. Unlike other indicators of cooperation between scientific and industrial research such as informal collaborations or researchers from the academic world working for industry, the NPL study shows explicit, voluntary, selective and asymmetric connections. This method makes it possible to study the A&D sector, which holds a special place in innovation systems.

### 2.3.2. *Data source and analysis*

#### 2.3.2.1. *The quality of patents*

This study of NPL in A&D patents must be complemented by an analysis of the influence of the diffusion of scientific knowledge on the quality of innovation. Indeed, from the knowledge-based economy perspective, it can be suggested that increased collaboration between different knowledge producers accelerates and improves the innovation process. Three main arguments can be used to explain how science contributes to innovation in industry:

– by avoiding irrelevant experiments and focusing on relevant research paths [FLE 04];

– by increasing and improving the firm's internal and external absorption capacity for new knowledge [COC 98];

– by improving the quality and economic value of technologies produced by companies [ARO 94].

In this chapter, we focus on the contribution of science to the quality of innovation in the A&D sector. The quality of a patent can be assessed using several criteria. The first criterion is the "commercial" value. In this case, the quality of a patent is treated as the economic value of the patent on the market, i.e. the increase in value of a firm's patent portfolio following the filing of the patent. On the contrary, the quality of a patent can be seen from a "technological" perspective. Here, the quality of the patent depends on a balance of generality and originality. In order to assess the quality, both from a commercial and technological perspective, we use four patent quality indicators, discussed in the OECD report [SQU 13].

First, the commercial quality of a patent can be estimated using the *size of the patent family*, known as *NPN*. Note that a group of patents protecting the same innovation is called a family of patents. Here, the value of the patent is associated with the geographical scope of innovation protection [LAN 98, HAR 03] show that a family of patents with representations in many countries is an indicator of the economic value of the patent. Indeed, extending the protection of an innovation to several countries around the world reflects, on the part of applicants, a willingness to accept patent costs as well as additional administrative delays. However, these additional costs are only incurred if the firm, or the applicant, considers that it will derive a higher benefit from them.

Second, the generality of a patent can be assessed by the *patent scope* corresponding to the number of different technological fields referenced in the patent family, noted *nbIPC*. Indeed, the patent examiner assigns to each patent one or more four-digit technological fields based on the International Patent Classification (WIPO). This indicator makes it possible to capture the scope of innovation and its influence on the market. [LER 94] observes that the technology contained in the patent portfolios of firms significantly affects the firm's market value. The idea behind this argument is that patents are more valuable when many substitutes in the same product class are available and they are owned by the same firm. This argument is all the more true when innovations potentially influence different industrial sectors.

Third, the originality of a patent can be assessed by the number of *backward citations* in the patent. In order to evaluate the novelty of an innovation, an inventor must cite all previous patents related to his invention. The patent examiner must ensure that all appropriate patents are cited. These make it possible to certify the patentability of the invention and define the legitimacy of the claim of paternity. [CRI 08] explain that the number of citations in a patent indicates the degree of novelty of the invention and evaluates the knowledge transfers made to produce that invention. In addition, aggregating patent citations at the technology level helps to shed light on the innovation process and its dynamics. [HAR 03] find a positive correlation between the number of patents cited and the market value of a patent. On the contrary, this indicator makes it possible to highlight the nature of the innovation process, i.e. whether the invention is incremental or disruptive. Thus, when the number of patents cited is small, it is easy to conclude that the technological invention concerned is not based on past innovations, and will therefore be more of a breakthrough. On the contrary, when the number of patents cited is large, it is interpreted as an invention based on a large number of past innovations, and will therefore be rather incremental in nature. In order to use such an indicator, it is essential to control the inventors' self-citations. Indeed, self-citations constitute both a bias due to strategic behavior on the part of the inventor – the inventor can self-cite in order to increase the quality statistics on these patents – as well as a knowledge bias – the inventor cites himself because that is what he knows. For this reason, we exclude self-citations.

Finally, forward citations reflect both the technological importance of the patent and the economic and commercial value of the invention [TRA 90]; [HAR 03]; [HAL 05]. The number of patents cited reflects the technological importance. The idea is that the innovation contained in the patent leads to a cluster of incremental innovations. The number of patents cited also indicates the economic value of a patent by indicating to the entire industrial sector the potential for innovation and thus increases the market value of the filing firm.

From this set of patent quality indicators, we aim to establish the link between patents and changes in innovation dynamics, reflected in the quantity of NPL contained in patents (noted *NPL/patent*).

#### 2.3.2.2. *Patents*

In order to assess the contribution of scientific knowledge to technological innovation, we use patents as a proxy for technological innovation in the A&D sector. As mentioned above, patents are good estimates of technological innovation [GAR 79]. Indeed, any organization that succeeds in seizing technological opportunities is confronted with the question of protecting its lead [TEE 88]. To this end, there are various possible protection regimes, including patents, which are particularly effective in claiming product innovation, and secrecy, which, in turn, would make it easier to assert one's rights over process innovations. The choice between these two methods of protecting innovation is made by the organization that created it. However, in the case of defence-related innovations, the State may force an organization to keep the invention a secret, even though it would have preferred to publish a patent. Indeed, with regard to sovereignty issues and despite the potential disclosure of a technology being considered high, the government is in a position to intervene to limit the economic exploitation of an innovation. This information bias created by the existence of secrecy as intellectual property protection is an unfortunate hindrance to the empirical analysis of innovation. Indeed, the very nature of this bias, secrecy, means that it cannot be detected. Hence, we use patents as an estimate of innovation, which like any indicator is imperfect but is also the most complete which researchers can access. It is a very informative document with data on both the nature of the innovation (technological field) and the knowledge obtained to produce it. In addition, this harmonized legal document, controlled and referenced by experts in technological fields, is more reliable and detailed than any other data on innovation.

Patents are industrial property rights that provide protection in one or more countries. Inventions are often the subject of several patent applications to different national or regional patent offices or to the World Intellectual Property Organization (WIPO). WIPO is the global forum for intellectual property services, policies, information and cooperation. Hence, analysis in terms of patent family is more relevant when looking at innovation. Indeed, the patent family makes it possible to locate a current innovation and not simply the reproduction of an innovation in a national or regional context. Orbit-Questel[3] has developed a definition of the family that combines the strict family rule of the European Patent Office (EPO)[4] with additional rules that take into account links with the parent EPO and/or PCT application, the links between provisional US applications and published US applications. This definition also takes into account their different definitions of invention.

To study the technologies contained in patents, we use the WIPO International Patent Classification, which is unanimously recognized and used for patent study. It divides the technology into eight sections with about 70,000 subdivisions. We use the patent database generated by Orbit. Between 1945 and 2015, we identified 51,161 patent families containing the A&D technology fields, filed in more than 63 filing offices worldwide (see Appendix 2 for the distribution of patent families by initial filing office).

3 Orbit-Questel is a database specialized in intellectual property, and in particular on patents.

4 EPO defines a patent family as including all documents with exactly the same priority or combination of priorities.

## 2.4. Results

In this section, we present the empirical results of our study. First, a descriptive analysis of patent family data in the A&D domain is developed, in particular an account of the use of scientific knowledge in these families. In a second step, we develop this descriptive study by introducing the correlations between patent quality and the NPL.

### 2.4.1. *Descriptive analysis*

The study of these patent families and their associated data makes it possible to highlight major trends. First, we observe a strong increase over time in patent family filings, which is in line with the literature (e.g. [KOR 99, KOR 00]; [HAL 01]; [HAL 04]; [JIN 04]). This trend is due to multiple factors: the entry of new countries into the "patent market", notably China and South Korea (WIPO, 2016) and [KOR 99] attribute it to the combination of the overall increase in innovation in society and the improvement of R&D management. Third, the increase in the importance of patent portfolios to protect against competitors has contributed to this trend [HAL 03]. Indeed, the increased value that financial markets attach to patent portfolios leads companies, especially new firms, to secure their technological innovations through patenting.

On the contrary, the referencing of scientific contributions to innovation contained in patents is also increasing sharply. As shown in Figure 2.1, the first reference to scientific literature in our patents came in the early 1970s. These references contain both articles published in peer-reviewed journals and scientific books as well as conference proceedings. This appearance of NPL in patents and its constant increase can be explained by three main elements. First, filing offices are increasing their requirements in terms of patent completeness. Second, the work of experts in filing offices responsible for patent referencing is becoming increasingly precise. Finally, our starting hypothesis which assumes that the change in the innovation model has led to a growth in technological innovations driven by an increased diffusion of scientific knowledge. As we have already presented, the rise of intangible capital together with the revolution of information and communication technologies has led to the development of a knowledge-based economy [FOR 98]. The resulting perpetual technological change is reflected in a multilateral, multidisciplinary and ever-changing combination of scientific and technological research and invention methods. These new trends are associated with an increase in the use of knowledge generated by university research. According to the OECD [OEC 96], the share of high technology in production, which is largely dependent on the production and dissemination of knowledge, particularly scientific knowledge, doubled between 1970 and 1990 in industrial production and exports, and this trend is still continuing today. This OECD finding is supported by our study of A&D technology sectors. Figure 2.1 shows a growing proportion of patents citing at least one scientific literature reference. The number of patents citing NPL was almost non-existent before the 1980s and increased to more than a third of patents from the 2000s onwards.

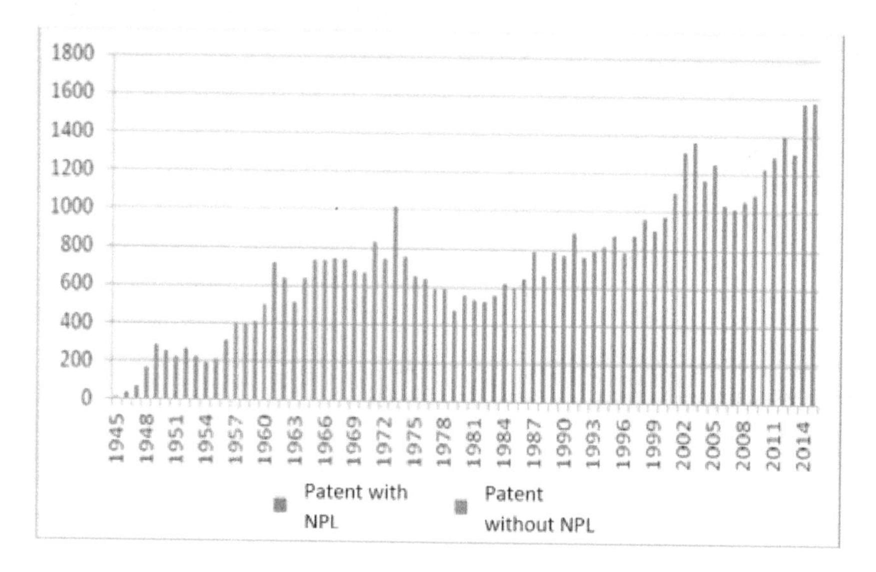


**Figure 2.1.** *The evolution of patents and the NPL of A&D between 1945 and 2015 (source: based on Orbit data). For a color version of the figures in this chapter see www.iste.co.uk/barbaroux/technology.zip*

A&D patents are composed of five distinct technology classes and their weight in the field changes over the period. Thus, Figure 2.2 shows that weapons technology (F41) dominates the A&D domain over the entire period accounting up to 60% over the last period (1996–2015). The increase in fraction is driven by armor and other means of attack or defence, such as camouflage. If the knowledge-based economy is an adequate analysis of our society, then this technological field, because of its expansion, should be intensive in the use of scientific knowledge. This is what we propose to study in Figure 2.3.

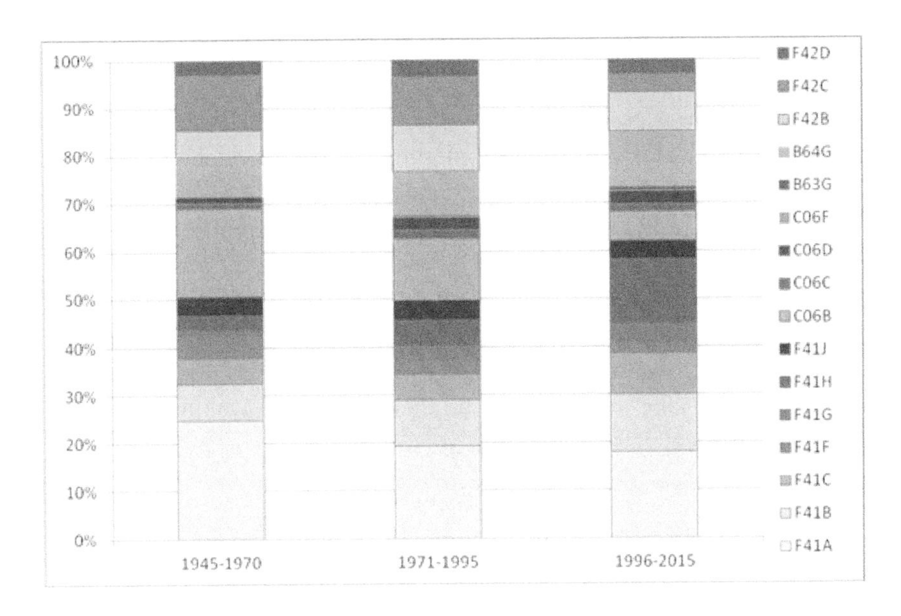


**Figure 2.2.** *Distribution of A&D patents by technology area between 1945 and 2015 (source: based on Orbit data)*

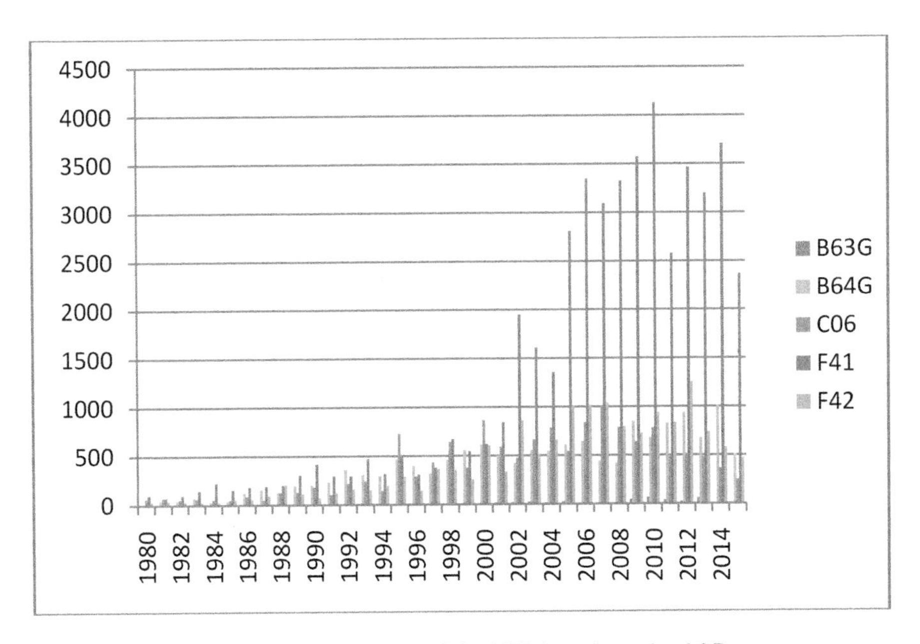


**Figure 2.3.** *Evolution of the NPL in volume by A&D technology area (source: based on Orbit data)*

In order to make the results more readable, we propose to analyze the technology classes grouped at the three-digit level for F41, F42 and C06. We present this analysis from 1980 onwards because it is from this period that the presence of NPL in patents emerges in all the technological fields studied. Figure 2.3 shows that the increase in the share of F41 in all patent families in the A&D domain is accompanied by a significant increase in NPL citations in this class. Indeed, even though the quantity of NPL increases in all technology classes over the period 1980–2015, it is the technologies relating to F41 that show the largest increase in volume of NPL, particularly from 2002 onwards. However, this study in terms of volume does not take into account the increase in NPL relative to the share of the technological class in the field of A&D.

Figure 2.4 shows the evolution of the NPL reference number in patents by A&D technology, in base 100 of the year 1980. We observe that all technologies have experienced strong growth in the NPL over the period 1980–2015. This growth is most significant for the B64G technology sector, corresponding to the technological innovation of vehicles and equipment for astronautics. This category includes all space vehicles, tools adapted to work in space, space clothing and simulators of living conditions in space. By means of illustration, one of the patents classified in this category is a satellite launcher, published in 1990 and citing a scientific article published in the *Journal of the British Interplanetary Society* in 1977 as NPL. The other technology areas studied have also experienced a significant increase in the amount of NPL in their patents, with an overall trend towards an acceleration of this increase. Indeed, between 1980 and 2010, the quantity of NPL in patents associated with offensive and defensive ship installations (B63G) increased by a factor of 60. We observe the same trend for weapons and ammunition technologies (F41 and F42 respectively). For innovation related to chemistry, and more particularly explosives (C06), unlike other areas, the growth of the NPL in patents is less erratic and increased eightfold between 1980 and 2015.

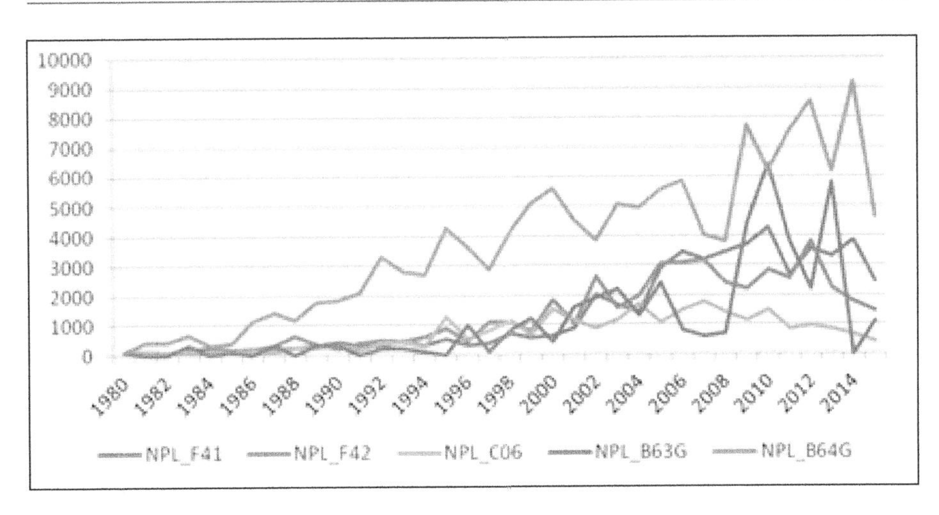


**Figure 2.4.** *Evolution of the NPL by A&D technology area (base 100: 1980) (source: based on Orbit data)*

Thus, this descriptive study of patent data and their citations from NPL highlights the simultaneous increase in the mobilization of scientific knowledge in industry and the production of technological innovation. However, this descriptive study is not sufficient to understand how the increased use of scientific knowledge in the production of technological innovation influences the ability to innovate, particularly in qualitative terms. To this end, we propose in the next section to study the link between the quantity of NPL in patent families and the quality of these same patents by technology class in the field of A&D.

### 2.4.2. *Scientific knowledge and quality of technological innovation*

Figure 2.5 shows the correlation tables between the quantity of NPL and patent quality indicators. The correlation coefficients (Pearson's) were calculated from 1980 to 2015. These correlation tables highlight several elements. Overall, we observe a positive and significant correlation between the quantity of NPL in patents and the various quality indicators, despite a low Pearson's coefficient (on average between 0.2 and 0.3) for the majority of observations. Thus, indicators of economic quality and technological generality are not very closely linked to the quantity of scientific knowledge mobilized. However, one correlation stands out particularly for all technology classes: the correlation between the quantity of NPL and patents cited (0.6). The contribution of scientific knowledge is therefore positively and significantly correlated with the originality of patents. Patents, based on a lot of scientific knowledge, are also those that include incremental inventions. This trend can be explained by two factors. On the one hand, the improvement in patent intelligence is both due to the increased vigilance of patent examiners and the growing role of intellectual property services in innovative firms. On the other hand, it highlights the emergence of increased collaborations between producers of scientific and technological knowledge. Indeed, we observe that an increase in the number of patents cited, reflecting the incremental nature of innovation, increases the quantity of NPL present in the same patent. As we have developed in the methodology section, a significant number of patents cited reflect the incremental nature of innovation. Thus, this correlation between *NPL/patent* and *NSCT* can be interpreted as the incrementality of A&D innovations being based not only on previous technological innovations but also on the scientific knowledge produced. It can be seen that this link is particularly strong for the technological field of weapons (F41).

If we look at the results by technological fields, we note that only the B63G technology, i.e. offensive or defensive installations on ships, submarines and aircraft carriers, has different coefficients than those in the overall table. Indeed, the correlation coefficient between the NPL/patent and the nbIPC, the size of the family and the number of patents cited is higher than for the other technological fields (between 0.3 and 0.4), while that between the NPL/patent and the number of patents cited is not significant. This technology is very specific because it is characterized by a strong military component and a rather limited field of action: the naval sector.

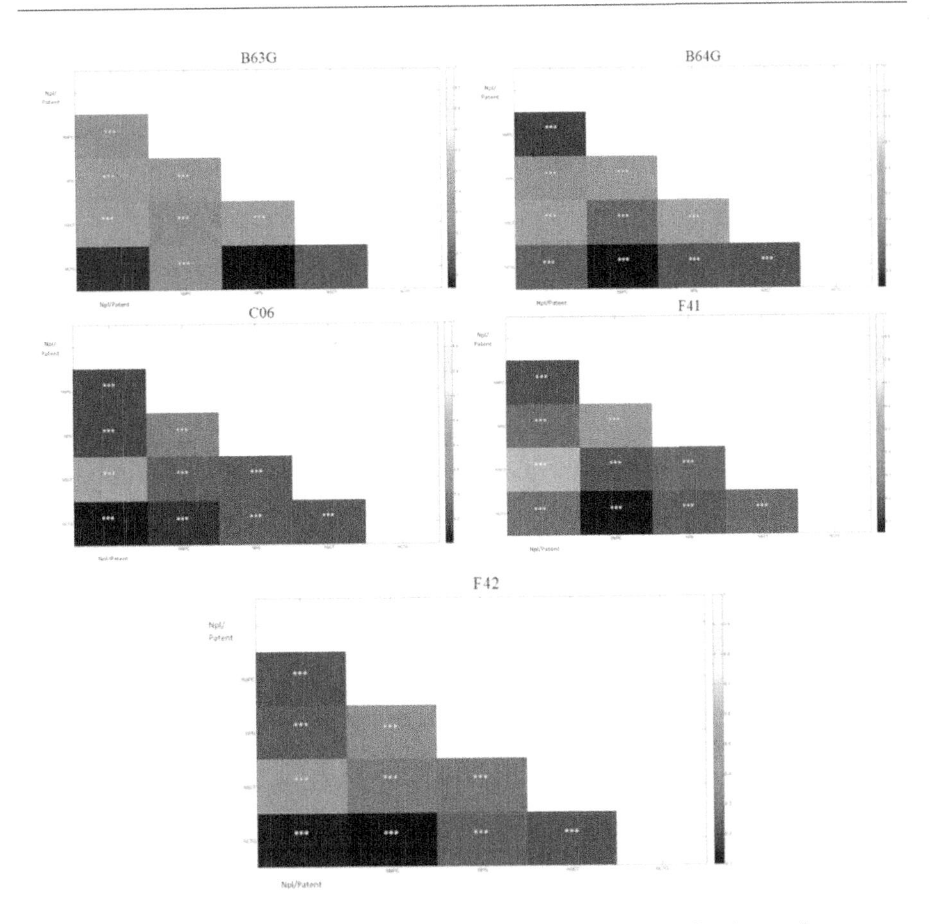


**Figure 2.5.** *Correlation tables between scientific contribution and patent quality between 1980 and 2015 by technological fields*

Thus, this study of correlation coefficients revealed a fairly weak, although significant, linear relationship between patent quality and the quantity of NPL present in patents. This confirms our intuition that the emergence of the knowledge-based economy is resulting in increased collaboration between producers of scientific and industrial knowledge, leading to increased production of technological innovation in terms of quality. In addition, this trend towards an increase in the NPL highlights a change in the nature of incremental A&D innovation, which is increasingly based on scientific knowledge.

## 2.5. Conclusion

In this chapter, we have assessed the changes in the innovation model of the A&D sector in terms of the contribution of scientific knowledge. First, we analyzed the evolution of the contribution of scientific knowledge to technological innovations in A&D between 1945 and 2015. This has shown an increase in the NPL present in patents, with the emergence of this type of citation from the 1980s onwards. This emergence corresponds to the beginning of what economists call the "knowledge-based economy". Thus, we can conclude that the A&D sector is no exception to the changes in innovation trends observed throughout the economy. This thus reinforces [COW 95]'s argument that questions the linear specificity of the military innovation model defended by [GAN 92]. Second, we have developed this subject through the study of patent quality and its link with the NPL. This analysis highlighted the co-evolution of patent and NPL citations throughout the A&D sector, as well as the particularity of the B63G technology area, encompassing offensive and defensive equipment for ships, submarines and aircraft carriers. It appears that, unlike the other technological classes of A&D, the incremental nature of inventions, i.e. their degree of generality, is not correlated with the mobilization of scientific knowledge. On the contrary, it is strongly linked to the economic value of inventions. However, our analysis is biased by the *a priori* selection of defence-related technological fields without addressing the question of the duality of such innovations. This quantitative study of innovation has thus highlighted the emergence of the knowledge-based economy, resulting in an increase in collaboration between producers of scientific and industrial knowledge. In the future, a similar study should be carried out by selecting not the A&D technological fields but firms with defence activities.

## 2.6. Appendices

| | |
|---|---|
| B63G | *Offensive or defensive arrangements on vessels, mine-laying, ...* |
| B64G | *Cosmonautics, vehicles or equipment therefor (apparatus for, or methods ...* |
| C06B | *Explosive or thermic compositions (blasting f42d), manufacture thereof, ...* |
| C06C | *Detonating or priming devices, fuses, chemical lighters, pyrophoric ...* |
| C06D | *Means for generating smoke or mist, gas-attack compositions, generation ...* |
| C06E | *Matches, manufacture of matches* |
| F41A | *Functional features or details common to both small arms and ordnance, ...* |
| F41B | *Weapons for projecting missiles without use of explosive or combustible ...* |

| | |
|---|---|
| F41C | *Small arms, e.g. Pistols, rifles (functional features or details common ...* |
| F41F | *Apparatus for launching projectiles or missiles from barrels, e.g....* |
| F41G | *Weapon sights, aiming (optical aspects thereof g02b)* |
| F41H | *Armor, armored turrets, armored or armed vehicles, means of attack or ...* |
| F41J | *Targets, target ranges, bullet catchers* |
| F42B | *Explosive charges, e.g. For blasting, fireworks, ammunition (explosive ...* |
| F42C | *Ammunition fuses (blasting cartridge initiators f42b0003100000, chemical ...* |
| F42D | *Blasting (fuses, e.g. Fuse cords, c06c0005000000, blasting cartridges ...* |

**Appendix 1.** *The A&D technology subclasses (OST, 2010)*

| First filing office | Number of patent families |
|---|---|
| Argentina | 12 |
| Austria | 159 |
| Australia | 321 |
| Belgium | 120 |
| Bulgaria | 5 |
| Brazil | 29 |
| Canada | 267 |
| Switzerland | 501 |
| Chile | 3 |
| China | 84 |
| Colombia | 5 |
| Costa Rica | 1 |
| Czechoslovakia | 21 |
| Cuba | 1 |
| Czech Republic | 26 |
| Germany | 3175 |
| Denmark | 38 |
| Eurasian Patent Organization | 2 |
| European Patent Office | 337 |
| Spain | 78 |
| Finland | 91 |
| France | 1961 |
| United Kingdom of Great Britain and Northern Ireland | 1244 |
| Greece | 9 |
| Hong Kong | 15 |
| Hungary | 11 |
| Ireland | 5 |

| Israel | 367 |
|---|---:|
| India | 17 |
| Italy | 409 |
| Japan | 1059 |
| Korea (the Republic of) | 249 |
| Luxembourg | 83 |
| Monaco | 1 |
| Mexico | 12 |
| Netherlands | 63 |
| Norway | 92 |
| New Zealand | 30 |
| Peru | 4 |
| Philippines | 5 |
| Poland | 10 |
| Portugal | 5 |
| Romania | 2 |
| Serbia | 1 |
| Russia (Federation) | 57 |
| Saudi Arabia | 1 |
| Sweden | 720 |
| Singapore | 26 |
| Slovenia | 1 |
| Slovakia | 1 |
| San Marino | 1 |
| Soviet Union (USSR) | 6 |
| Thailand | 1 |
| Turkey | 11 |
| Taiwan (Province of China) | 120 |
| Ukraine | 9 |
| United States of America | 38794 |
| Uruguay | 1 |
| Venezuela (Bolivarian Republic of) | 2 |
| World Intellectual Property Organization (WIPO) | 277 |
| Yugoslavia/Serbia and Montenegro | 1 |
| South Africa | 204 |
| Zimbabwe | 1 |

**Appendix 2.** *Distribution of patent families by first filing office*

## 2.7. References

[ALB 91] ALBERT, M.B., AVERY, F., NARIN, F., MCALLISTER P., 1991. "Direct validation of citation counts as indicators of industrially important patents", *Research Policy*, 20 (3): 251-59.

[ARO 94] ARORA, A., GAMBARDELLA, A., 1994. "The changing technology of technological change: General and abstract knowledge and the division of innovative labour", *Research Policy, Special Issue in Honor of Nathan Rosenberg*, 23 (5): 523-32.

[BEL 17] BELLAIS, R., DROFF, J. 2017. "L'innovation technologique tient une place essentielle dans la défense depuis le XXe siècle. Le modèle d'innovation pour les équipements militaires est fortement marqué par l'héritage conceptuel issu de la Guerre..." *Annuaire Français de Relations Internationales*, vol. 18.

[BRA 05] BRANSTETTER, L., OGURA, Y., 2005. "Is academic science driving a surge in industrial innovation? Evidence from patent citations", *Working Paper 11561*, National Bureau of Economic Research. Available at: http://www.nber.org/papers/w11561.

[BRA 74] BRAVERMAN, H., 1974. *Labor and Monopoly Capital: The Degradation of Work in the Twentieth Century*, Monthly Review Press New York.

[BRE 02] BRET, P., 2002. "L'État, l'armée, la science. L'invention de la recherche publique en France (1763–1830)". *Annales historiques de la Révolution française*, 328 (1): 278-278.

[CAR 81] CARPENTER, M.P., NARIN, F., WOOLF, P., 1981. "Citation rates to technologically important patents", *World Patent Information*, 3 (4): 160-63.

[COC 98] COCKBURN, I.M., HENDERSON, R.M., 1998. "Absorptive capacity, coauthoring behavior, and the organization of research in drug discovery", *The Journal of Industrial Economics*, 46 (2): 157-82.

[COH 02] COHEN, W.M., NELSON, R.R., WALSH, J.P., 2002. "Links and impacts: The influence of public research on industrial R&D", *Management Science*, 48 (1): 1-23.

[COO 98] COOMBS, R., HULL, R., 1998. "Knowledge Management Practices and Path-Dependency in Innovation", 27(3): 237-253.

[COW 95] COWAN, R., FORAY, D., 1995. "The changing economics of technological learning", *Monograph*, May 1995. Available at: http://pure.iiasa.ac.at/4553/.

[CRI 08] CRISCUOLO, P., VERSPAGEN, B., 2008. "Does it matter where patent citations come from? Inventor vs. examiner citations in European patents", *Research Policy, Special Section Knowledge Dynamics out of Balance: Knowledge Biased*, Skewed and Unmatched, 37 (10): 1892-1908.

[DEP 14] DEPEYRE, C., 2014. "Boeing boeing : La dualité civil-militaire source d'un rebond stratégique dans l'ère post-guerre froide, "Boeing Boeing": The civil-military balance, source of a strategic bounce in the post cold war era", *Entreprises et histoire*, no. 73 (May): 58-74.

[DIT 06] DITTRICH, K., FERDINAND, J., VAN DER VALK, W., WYNSTRA, F., 2006. "Dealing with dualities", *Industrial Marketing Management*, IMP 2005: Dealing with Dualities, 35 (7): 792-96.

[DUP 13] DUPUY, R., 2013. "L'industrie européenne de défense: Changements institutionnels et stratégies de coopétition des firmes", *Innovations*, 42 (3): 85. Available at: https://doi.org/10.3917/inno.042.0085.

[ETZ 00] ETZKOWITZ, H., LEYDESDORFF, L., 2000. "The dynamics of innovation: From national systems and "Mode 2" to a Triple Helix of university–industry–government relations", *Research Policy*, 29 (2): 109-23.

[FAU 16] FAUCONNET, C., MALIZARD, J., 2016. "Rétrospective des exportations d'armement en France (1958–2015)". *Revue Défense Nationale*.

[FLE 04] FLEMING, L., SORENSON, O., 2004. "Science as a map in technological search", *Strategic Management Journal*, 25 (8-9): 909-28.

[FOR 98] FORAY, D., LUNDVALL, B., 1998. "The knowledge-based economy: From economics of knowledge to the learning economy", in *The Economic Impact of Knowledge*, Routledge.

[GAN 92] GANSLER, J.S., 1992. "Restructuring the defence: Industrial base", *Issues in Science and Technology*, 8 (3): 50-58.

[HAL 04] HALL, B.H., 2004. "Exploring the patent explosion", *The Journal of Technology Transfer*, 30 (1-2): 35-48.

[HAL 05] HALL, B.H., JAFFE A., TRAJTENBERG M., 2005. "Market value and patent citations", *The RAND Journal of Economics*, 36 (1): 16-38.

[HAL 01] HALL, B.H., ZIEDONIS, R.H., 2001. "The patent paradox revisited: An empirical study of patenting in the U.S. semiconductor industry, 1979-1995", *The RAND Journal of Economics*, 32 (1): 101-28.

[HAR 03] HARHOFF, D., SCHERER, F.M., VOPEL, K., 2003. "Citations, family size, opposition and the value of patent rights", *Research Policy*, 32 (8): 1343-63.

[HIC 01] HICKS, D., BREITZMAN, T., OLIVASTRO, D., HAMILTON, K., 2001. "The changing composition of innovative activity in the US — a portrait based on patent analysis", *Research Policy*, 30 (4): 681-703.

[HOB 98] HOBDAY, M., 1998. "Product complexity, innovation and industrial organisation", *Research Policy*, 26 (6): 689-710.

[KEL 95] KELLEY, M.R., WATKINS, T.A., 1995. "In from the cold: Prospects for conversion of the defence industrial base", *Science*, 268 (5210): 525-32.

[JIN 04] JINYOUNG, K., MARSCHKE, G., 2004. "Accounting for the recent surge in U.S. patenting: Changes in R&D expenditures, patent yields, and the high tech sector", *Economics of Innovation and New Technology*, 13 (6): 543-58.

[KOR 99] KORTUM, S., LERNER, J., 1999. "What is behind the recent surge in patenting? The paper is a condensed version of 'Stronger Protection or Technological Revolution: What is Behind the Recent Surge in Patenting?' Carnegie-Rochester Conference Series on Public Policy, vol. 48, 1998, 247–304", *Research Policy*, 28 (1): 1-22.

[KOR 00] KORTUM, S., LERNER, J., 2000. "Assessing the contribution of venture capital to innovation", *The RAND Journal of Economics*, 31 (4) (Winter): 674-692.

[LAN 98] LANJOUW, J.O., PAKES, A., PUTMAN, J., 1998. "How to count patents and value intellectual property: The uses of patent renewal and application data", *The Journal of Industrial Economics*, 46 (4): 405-32.

[LEN 02] LENFLE, S., MIDLER, C., 2002. "Stratégie d'innovation et organisation de la conception dans les entreprises amont", *Revue Française de Gestion*, 28 (140): 89-105.

[LER 94] LERNER, J., 1994. "The importance of patent scope: An empirical analysis", *The RAND Journal of Economics*, 25 (2): 319-33.

[LEY 95] LEYDESDORFF, L., 1995. "The triple Helix—University-Industry-Government Relations: A laboratory for knowledge-based economic development", *EASST Review*, 14-19.

[LOO 03] LOOY, B., VAN DEBACKERE, K., ANDIRES, P., 2003. "Policies to stimulate regional innovation capabilities via University–industry collaboration: An analysis and an assessment", *R&D Management*, 33 (2): 209-29.

[MIN 15] MINGERS, J., LEYDESDORFF, L., 2015. "A review of theory and practice in scientometrics", *European Journal of Operational Research*, 246 (1): 1-19.

[MOW 01] MOWERY, D.C., NELSON, R.R., SAMPAT, B.N., ZIEDONIS, A.A., 2001. "The growth of patenting and licensing by U.S. Universities: An assessment of the effects of the Bayh–Dole act of 1980", *Research Policy*, 30 (1): 99-119.

[NAR 85] NARIN, F., NOMA, E., 1985. "Is technology becoming science?", *Scientometrics*, 7 (3-6): 369-81.

[NAR 97] NARIN, F., HAMILTON, K.S., OLIVASTRO, D., 1997. "The increasing linkage between U.S. technology and public science", *Research Policy*, 26 (3): 317-30. Available at: https://doi.org/10.1016/S0048-7333(97)00013-9.

[NEL 85] NELSON, R.R., WINTER, S.G., 1985. *An Evolutionary Theory of Economic Change*, Belknap Press, Cambridge.

[NOB 79] NOBLE, D.F., 1979. *America by Design: Science, Technology, and the Rise of Corporate Capitalism*, Oxford University Press, Oxford.

[OST 10] OBSERVATOIRE DES SCIENCES ET TECHNIQUES (OST), 2010. "Indicateurs de sciences et de technologies", *Les rapports et les analyses de l'OST*. Economica.

[OEC 96] OECD, 1996. "The knowledge-based economy - OECD". Available at: http://www.oecd.org/sti/sci-tech/theknowledge-basedeconomy.htm.

[OEC 05] OECD, 2005. "Manuel d'Oslo: Principes directeurs pour le recueil et l'interprétation des données sur l'innovation, 3e édition", Available at: http://www.oecd.org/fr/science/inno/manueldosloprincipesdirecteurspourlerecueiletlinterpretationdesdonneessurlinnovation3eedition.htm.

[SAL 01] SALTER, A.J., MARTIN, B.R., 2001. "The economic benefits of publicly funded basic research: A critical review", *Research Policy*, 30 (3): 509–32.

[SAM 03] SAMPAT, B.N., MOWERY, D.C., ZIEDONIS, A.A., 2003. "Changes in university patent quality after the Bayh–Dole act: A re-examination", *International Journal of Industrial Organization*, The Economics of Intellectual Property at Universities, 21 (9): 1371-90.

[SEM 15] SEMPERE, C.M., 2015. "A survey of performance issues in defence innovation", *Defence and Peace Economics*, 28 (3): 319-43.

[SQI 13] SQUICCIARINI, M., DERNIS, H., CRISCUOLO, C., 2013. "Measuring patent quality: Indicators of technological and economic value", *OECD*.

[TEE 88] TEECE, D.J., 1988. "Capturing value from technological innovation: Integration, strategic partnering, and licensing decisions", *Interfaces*, 18 (3): 46-61.

[TIJ 04] TIJSSEN, R.J.W., 2004. "Measuring and evaluating science—technology connections and interactions", in MOED, H.F., GLÄNZEL, W., SCHMOCH U. (eds), *Handbook of Quantitative Science and Technology Research*, 695-715. Springer Netherlands.

[TRA 90] TRAJTENBERG, M., 1990. "A penny for your quotes: Patent citations and the value of innovations", *The RAND Journal of Economics*, 21 (1): 172-87.

[VER 03] VERBEEK, A., DEBACKERE, K., LUWEL, M., 2003. "Science cited in patents: A geographic "flow" analysis of bibliographic citation patterns in patents", *Scientometrics*, 58 (2): 241-63.

[WIP 17] WIPO, "Statistiques de propriété intellectuelle", Accessed 26 September 2017. Available at: /ipstats/fr/index.html.

# 3

# Identification of Defence Technological Knowledge Systems: A Tool for Duality Analysis

ABSTRACT. The purpose of this chapter is to offer a method for empirically identifying the knowledge systems that structure the "technological landscape" within the defence industries. In order to do this, this chapter proposes shedding light on Technological Knowledge Systems. Isolating these systems makes it possible to shed light on not only component knowledge but also architectural knowledge, which, for each system, structures innovation dynamics in the defence sector.

## 3.1. Introduction

Since the 1980s, an increase in debates on technological transfers between the civil and defence sectors (Alic *et al.*, 1992) and the development of the concept of duality (e.g. Gummet and Reppy, 1988) have made the gap between military and civil technology more tenuous. The dynamics of innovation within the defence technological and industrial base (DTIB) are now largely influenced by those within civil industries (Meunier, 2017). Identifying the dynamics of innovation specific to the defence sector within the DTIB has become more complex. However, despite this proximity, identifying these dynamics remains a challenge for public authorities and industrialists. They have not only to identify the technologies needed to develop today's defence systems, but also ensure that the skills and

---

Chapter written by François-Xavier MEUNIER.

knowledge essential to tomorrow's systems are maintained. Moreover, good understanding of the defence technological perimeter and its characteristics is necessary for the integration of defence innovation into an increasingly dual and globalized technological perimeter.

The objective of this chapter is to propose a method for empirically identifying the knowledge systems that structure the defence technology field. To do this, this chapter proposes to highlight technological knowledge systems (TKS). Isolating TKSs makes it possible to highlight not only the component knowledge but also the architectural knowledge which, for each system, structure the dynamics of innovation in the defence field.

First, this chapter will return to the interest of identifying knowledge systems at the heart of defence innovation. Then, the identification method will be detailed before presenting the results obtained.

## 3.2. Definition of a TKS and defence innovation

This chapter proposes an original method for analyzing the knowledge base of firms. It is based on the concept, defined in this section, of "technological knowledge systems" (TKSs) and takes advantage of the information contained in patent applications. The aim of this method is to emancipate itself from the case study, commonly used to empirically study technological fields in the defence sector. To do this, the method is based on the study of knowledge that is a basic element of technological innovation (Carlsson and Stankiewicz, 1991).

Technological systems are approached from a systemic perspective in order to better understand the multiple modalities of technological interactions between the civil and military sectors. Indeed, it is by considering technological systems as a whole that we are able to measure the complexity of the knowledge included in the "technological system".

A "technological system" was defined by Carlsson and Stankiewicz: "*A technological system may be defined as a network of agents interacting in a specific economic/industrial area under a particular institutional infrastructure or set of infrastructures and involved in the generation, diffusion, and utilization of technology. Technological systems are defined in terms of knowledge/competence flows rather than flows of ordinary goods and services*" (Carlsson and Stankiewicz, p. 111, 1991). Technological systems are composed of four main elements:

– a hard core of technical and scientific knowledge;

– a constellation of technical systems;

– a market environment;

– an institutional interface.

This representation of the technological system seems entirely appropriate for a better understanding of the integration of defence into a broader technological environment. Indeed, it shows that the same knowledge base can lead to the development of different products that can therefore go beyond the sphere of defence or that the same product can require a combination of knowledge available beyond the sphere in which it has emerged. It shows that different products can address different consumers while they are based on technologies built on a base of similar technical and scientific knowledge. Thus, in this case, either one (or more) technology (s) may result in several distinct products, each of which is intended for military or civilian use; or a set of technologies from civil and military spheres may be used to develop a single product for one or both military and civilian use. In both cases, it is a body of knowledge that, combined in a way, makes it possible to bridge the gap between the civilian and military fields.

A related but distinct concept is a technical system (Gille, 1978). Nor does it focus only on the technical characteristics or sectors to which a system is attached, but more particularly seeks to understand how the elements of a system (actors, organizations, knowledge, applications, uses) are organized. In his analysis, he highlights the role of generic technologies, also known as General Purpose Technologies, (Bresnahan and Trajtenberg, 1995) in the deployment of the system (De Bandt, 2002). In the technical system, all the pieces of knowledge and technological skills are thus linked and made coherent at a functional level (Bainée, 2014).

Using the typology, between component knowledge and architectural knowledge (Henderson and Clark, 1990), identifying a technological system requires distinguishing, on the one hand, the knowledge bricks (component knowledge) that make up that system. They correspond to the units of knowledge, more or less important, that are associated within the system.

On the other hand, architectural knowledge corresponds to the way in which knowledge bricks are articulated between them and thus reveal the links that exist between the different bricks.

In order to observe this knowledge, two logics are opposed. The first is to identify systems from the output, i.e. the technological artifacts associated with the system and to find out what knowledge supports it. The other way of proceeding is based on knowledge. It consists of analyzing the interactions between knowledge and then identifying the applications and possibly artifacts that are the result of these interactions.

The first method does not seem appropriate for defence work. Indeed, for reasons of sovereignty, defence artifacts are often difficult to access and their technological characteristics even more so. Data are by definition sensitive; knowing in detail the characteristics of a particular weapon system in order to link it to a broader technical system poses obvious security problems if the information is disseminated (Buesa, 2001).

The methodology developed here is based on the second logic which seems more appropriate. It is based on knowledge analysis to delimit technological systems. This implies that exhaustive knowledge of military equipment is not necessary and takes advantage of technological knowledge data, considered through patents, much easier for an economist to collect and analyze. In addition, since the aim here is to analyze the scope of defence innovation in order to facilitate its integration into the global technological field (duality), sovereignty issues that could raise questions about the relevance of patent use are being eliminated. Indeed, for sovereignty and sensitive technologies, States ensure, with varying degrees of success, that these technologies do not go beyond the scope their control and that manufacturers keep this secret. For other technologies, the question of patenting is, as in any industry, a matter of business strategy. Therefore, the approach proposed here consists of studying the interactions between knowledge in order to determine which ones make up the system and make it possible to produce technological innovations.

It is for this purpose that the concept of technological knowledge system (TKS) is proposed. It is defined by analogy with a system of scientific knowledge that corresponds to the body of knowledge closely linked to each other, enabling the development and understanding of a scientific problem. For example, Cournot describes mathematics as a system of scientific knowledge "based on ideal notions that are in everyone's mind" (Cournot, 1847).

In this spirit, a TKS corresponds to all the knowledge that, closely linked to each other, creates synergies in technological production. This coherent body of knowledge (potentially based on scientific principles) combined with specific skills makes it possible to propose technical solutions that are then combined within one or more technological systems.

In order to represent and analyze the knowledge of defence companies, the knowledge base concept, as defined by Nesta and Saviotti (2005, 2006) or Krafft, Quatraro, Saviotti (2011), is used. A knowledge base corresponds to the network of patent technological classes that one or more organizations file. In the case of this chapter, companies with defence sales. This network or graph is characterized by the adjacency matrix whose characteristics will be studied later.

### 3.3. Data

The use of patent data has already demonstrated its relevance in the study of knowledge (Jaffe Trajtenberg, 2002; Verspagen, 2004). Leydesdorff *et al.* (2014) point out that patents provide a rich source of information on innovation activity and its context, specifying in particular the names of inventors and operators, the sources of their knowledge or the places where innovation activity takes place. The use of structural indicators makes it possible to analyze the production of knowledge and its dissemination using this information. Like many authors, this paper contributes to the necessary development of methods to study these data.

The patent citations allow the construction of technological flow matrices (Verspagent, 2004). These matrices are then used to study technological production. It is in order to make the best use of patent data, being aware of the associated limitations, that the method proposed here is based on building a knowledge base by considering a set of patents from companies with a defence activity. Therefore, this chapter focuses on formal knowledge. The formal part of knowledge is relatively easy to observe since by definition there is a trace of it through patents, for example; the tacit part is much less so.

However, if tacit knowledge is not materialized as such, by observing the organization of formalized knowledge through the knowledge architecture linking component knowledge between them in various achievements, it is possible to reveal the underlying informal part. Indeed, it is partly thanks to this informal part that formalized knowledge can be associated to produce a technology. Therefore, a better understanding of the structuring of knowledge is a challenge, not only for companies themselves, but also for all stakeholders who seek to understand the process of knowledge production and dissemination (public decision-makers, economists, regulators, etc.). This formalized knowledge can be observed at different levels: at the technological system level, organizational level, industry level and more specifically in this case at the knowledge base level.

This chapter proposes defence TKS identification based on the EC-JRC/OECD COR&DIP database, which lists the patent filings of 2000 groups, representing approximately 90% of private Research and Development (R&D) spending worldwide in 2012. We have decided to retain only patents filed at the European Patent Office (EPO), the American Patent Office (USPTO) or in one of the offices of the European Union 15 countries (Belgium, France, Germany, Italy, Luxembourg, the Netherlands, Denmark, Ireland, the UK, Greece, Spain, Portugal, Austria, Finland and Sweden).

As highlighted in the previous section, obtaining the list of knowledge contained in a defence technology system poses several problems. One of them is the dual nature of defence companies. Indeed, these companies are not purely defence and the share of their activity related to this field varies very strongly from one company to another. The choice that has been made is to focus the analysis on companies that, within the COR&DIP database, are active in the defence area and to consider the contribution of these companies to the defence knowledge base according to their activity in this field. These companies are identified using SIPRI data[1] (top 100 companies data in a time series from 2002 to 2015). This database lists the top 100 companies active in the defence sector each year. Between 2010 and 2012, 65 companies belonged to both the top 100 SIPRIs and the COR&DIP database. These companies account for 368,254 patents filed. This dataset focuses on large companies that supply defence equipment but which may be fairly diversified and belong to different industrial sectors. Another option

1 Stockholm International Peace Research Institute (SIPRI).

would have been to focus the analysis with a more limited scope by excluding, for example, companies whose main sector of activity is not in defence. Our option provides a broader overview of defence innovation.

Nevertheless, in order to capture the specificity of these companies' innovation activities in the defence area, technological flows within a company, whose contribution to defence is minimal, cannot be considered on the same scale as the flows observed in a company where defence represents all (or almost all) of its activity. Therefore, like Piscitello in his analyses on the diversification of firms, the results of the analysis of firms' knowledge bases are weighted by the share of the defence activity in the company's sales (Piscitello, 2005). This makes it possible to give greater value to technological innovations made in a company where defence activity is significant and makes it possible to highlight the defence nature of architectural and component knowledge specific to defence activities within companies' knowledge bases.

To respect this principle, the value of the coefficients of the technology flow matrix for a company is considered as a "layer". Each of the layers representing a company's knowledge base is weighted by the percentage of each company's sales related to the defence activity. To this end, the data published by SIPRI make it possible to know this defence-related sales among the 100 largest defence manufacturers and thus to give a different weight to these companies.

The resulting matrix then constitutes a portion of the overall input–output system corresponding to the proportion of defence within these technological flows (patent citations). It represents defence technological flows and therefore the defence knowledge base within which TKS are identified.

This matrix is interpreted as the adjacency matrix of the graph that traces the relationships between the different technological classes in the defence knowledge base. By using the typology between knowledge blocks and architectural knowledge (Henderson and Clark, 1990), such a graph distinguishes between component knowledge that correspond to technological classes and architectural knowledge that are represented by the links between technological classes.

### 3.4. Methodology

The method is anchored in the framework of the economic dominance theory (EDT) (Perroux, 1948) and mobilizes the tools of the influence graph theory (IGT) that develop from it (Lantner, 1974). In order to measure synergies in the production of technological knowledge, we structure the data in the form of an input–output (IO) system using citations between patents (Jaffe, Trajetenberg, 2002). This system takes the form of a graph and a square matrix containing the 641 technological classes of the International Patent Classification (IPC). By considering each company in proportion to its defence sales, it is possible to isolate the flows of knowledge related to this field. It is with this data that the defence knowledge base is built to measure the interdependencies between the technological classes (the poles of the graph) in order to highlight synergies and thus define TKS.

As Lebert and Younsi (2015) point out, the economic domination theory (EDT) is based on two foundations, one theoretical and the other methodological. Theoretically, the EDT is in line with the work carried out by F. Perroux on power relations (of dominance) in economics. These studies propose a notion of the economic world, as a set of relationships of domination that can be apparent or hidden (Perroux, 1948). Methodologically, it is related to graph theory in mathematics, which is the main tool of sociometric studies, also known as social network analysis (SNA). L.C. Freeman points out that the "Sorbonne School" provided an initial overview of these, "explicitly showing that a wide range of social problems could be understood as particular cases of a general structural model" (Freeman, p. 114, 2004).

Until now, the EDT has mainly been applied to the analysis of inter-industry relationships. First, from a national perspective, to study the relationships of dependence and interdependence between productive sectors as represented by national accounting (Lantner 1974) and, second, from an international perspective, to study the relationships of domination in international trade (Lebert and El Younsi, 2015). Based on the EDT, this chapter proposes to analyze knowledge production in order to understand the relationships of dependence and interdependence that structure it. For this purpose, the defence knowledge base is defined using patent data.

The use of patent citation data makes it possible to construct technological flow matrices (see Verspagen, 2004 for more details on the construction of flow matrices). These patent citations make it possible to analyze intertechnological relationships during the innovation process (Autant-Bernard *et al.*, 2014). The objective to draw a directed and weighted link between "citing" technology classes (those to which the "citing" patents are attached) and the cited technology classes (those to which the cited patents are attached). In addition to the weighting linked to defence sales described above, we specify three weights that are applied during the construction of the matrices.

First, the internal weighting of the patent. The technological links between the cited and citing classes are counted as a fraction of the number of possible combinations taking into account the number of technological classes referenced in each patent.

Second, the weighting linked to the companies holding the patent. If the patent in question is held by only one company, then it is not affected by this weighting. On the contrary, if several companies share ownership of a patent, then the value of the technological flows associated with it is distributed fairly among the various companies (fractional counting).

Third, the weighting related to patent families. As this study involves several intellectual property protection offices, we do not simply observe patents, but more precisely patent families. These families make it possible to group together several patents filed in different offices but referring to the same invention. Indeed, the same invention may be the subject of a patent in the USA, another in Europe, another specifically for the French Office, etc. Each patent then lists a list of "citing" and "cited" technology classes (TCs). However, each office is more or less independent and proceeds according to its own methods to identify the "citing" TC as well as the patents cited and therefore the "cited" TC for each patent (Mérindol and Fortune, 2009). It leads, from one patent to another, to gaps in the list of technological classes, even though they belong to the same family. Therefore, when aggregating these patents, the value of technological links should not be overestimated or underestimated. For this reason, we only take into account, as commonly done, the unique links between the TC of patents from the same family (e.g. an American patent citing a German patent and a French patent of the same family and with the same TC, combine identical citations which must be counted only once). We consider that the duplicates refer to the same technological links.

The benefit of analyzing technology flow matrices using the EDT is the identification of links between knowledge allowing the identification of synergies (interdependencies) and finally TKS within a knowledge base.

The principle of the influence graph theory (IGT) is to consider an input–output system whose Leontief matrix or Gosh matrix is calculated; the value of the interdependencies within the structure is equal to $I_G = 1 - D$, where $I_G$ is the general interdependence and D is the determinant of the structure. In order to respect the theorems of the IGT, the links with the outside and the self-consumption of the poles are considered as null (Lantner and Lebert, 2015). Here the objective is to analyze the interdependence between the poles, these precautions have no impact on the results while simplifying the calculations.

In order to measure the interdependencies between the poles of a structure using the IGT, a method has been previously developed (Lebert *et al.*, 2009); it consists of a robustness analysis whose principle is to proceed by iteration. The idea is, starting from the complete structure, to remove the arc on each turn, which once detached from the structure, maximizes the overall interdependence of the remaining structure. Thus, the first arc removed is the one that contributes the least to the overall interdependence, while the last one removed is the one that contributes the most. By classifying these arcs from the last to the first, it is possible to rebuild a structure that highlights the strongest relationships between the poles, which leads to the emergence of clusters.

Nevertheless, this procedure is at the same time long, tedious and consumes a lot of IT resources. Indeed, at each iteration, it is necessary to test the deletion of all arcs in order to define which one should be deleted. The methodological challenge is to propose here an alternative procedure to identify interdependencies between the poles in order to identify TKSs.

To do this, within a structure, where autarchy is null, a partition P is defined such that: $p_1$ is a main part made up of two poles and $p_2$ is a complementary part consisting of the rest of the n-2 pole structure. Then, for each of the possible partitions *P*, it can be shown that the sum of the intra- and inter-part interdependencies is greater than or equal to the general interdependence of the structure. Indeed, with the general interdependence $I_G = 1 - D$ and according to the partition theorem (Lantner and Lebert, 2013), the sum of intra- and inter-part interdependencies is equal to:

$$\sum_i^2 (1 - D_{pi}) + D_{p1}D_{p2} - D$$
$$= 1 - D_{p1} - D_{p2} + D_{p1}D_{p2} + (1 - D)$$
$$= \prod_i^2 \left(1 - D_{pi}\right) + I_G$$

However, given that $(1 - D_{pi}) \geq 0$, then:

$$\prod_i^2 \left(1 - D_{pi}\right) \geq 0$$

Therefore, with $I_G \geq 0$:

$$\prod_i^2 \left(1 - D_{pi}\right) + I_G \geq 0$$

The surplus with respect to $I_G$ measures synergy $S_P$ associated with a certain partition *P* and corresponds to the weight of internalized interdependencies in both parts:

$$S_P = \prod_i^2 \left(1 - D_{pi}\right)$$

Let us consider two cases:

If we propose the case where, the interdependencies within the complementary part are null, the sum of the interdependencies in the main part $(1 - D_{p1})$ and the interdependencies between the parties $(D_{p1} - D)$ is:

$$(1 - D_{p1}) + (D_{p1} - D) = I_G$$

Similarly, if there is no interdependence within the complementary part, then the sum of the interdependencies in the complementary part $(1 - D_{p2})$ and the interdependencies between the parts $(D_{p2} - D)$ is:

$$(1 - D_{p2}) + (D_{p2} - D) = I_G$$

In these two cases, there is no synergy related to the proposed partition, and therefore no surplus:

$$S_P = 0$$

A synergy $S_P > 0$ corresponds to a situation where, for a partition *P*, there are simultaneous interdependencies within the two parts. Finding the strongest synergies $S_P$ is finding the strongest combinations of $1 - D_{p1}$ and $1 - D_{p2}$, i.e. the interdependencies in the main part with regard to the values of the interdependencies in the complementary part.

The aim of the method is to calculate, for each of the combinations of both poles, the synergies they generate by taking into account both the interdependencies between these poles and the interdependencies in the rest of the structure. Thus, all other things being equal, a combination of two poles that undermines too strongly the synergetic potential of the rest of the structure (i.e. that greatly reduces the interdependence score of the complementary part) will obtain a lower synergy score than another combination that preserves more circularity in the rest of the structure. This allows the highest synergy scores to be assigned to combinations of two most interdependent poles that preserve circularity in the complementary part. This defines strong synergies between the two poles and a preservation of the synergetic potential in the rest of the structure. It is this synergy score linked to a particular partition that makes it possible to calculate a synergy score two by two between the poles of a structure.

To apply this method to the definition of TKS, it is sufficient to apply it to the matrix of technological coefficients in the defence knowledge base. Then, calculate for all combinations of technology classes the synergy scores as they have just been defined.

### 3.5. Results

In order to consider only the technologies that are within the scope of our structure and thus avoid taking into account, to an excessive extent, the synergies associated with General Purpose technologies, for example (Bresnahan and Trajtenberg, 1995), we have decided on a cut-off threshold.

For this purpose, the interdependence scores obtained are normalized between "0" and "1"; a score of "0" represents the strongest interdependence between two poles within the structure and a score of "1" represents the absence of interdependence. This results in relative synergy scores between two poles. Thus, only those interdependencies are considered whose scores are less than a certain percentage of the maximum score (the lowest interdependence). For example, by choosing a 35% cut-off threshold, only synergies whose score is less than 35% of the highest score are selected.

These relative synergy scores are classified within a symmetrical matrix. It is interpreted as a distance matrix with diagonal elements equal to 0 and is graphically represented using a dendrogram with a fraction in Figure 3.1. It is by studying this dendrogram, which represents the synergies between the technological classes of the defence knowledge base, that TKS are identified.

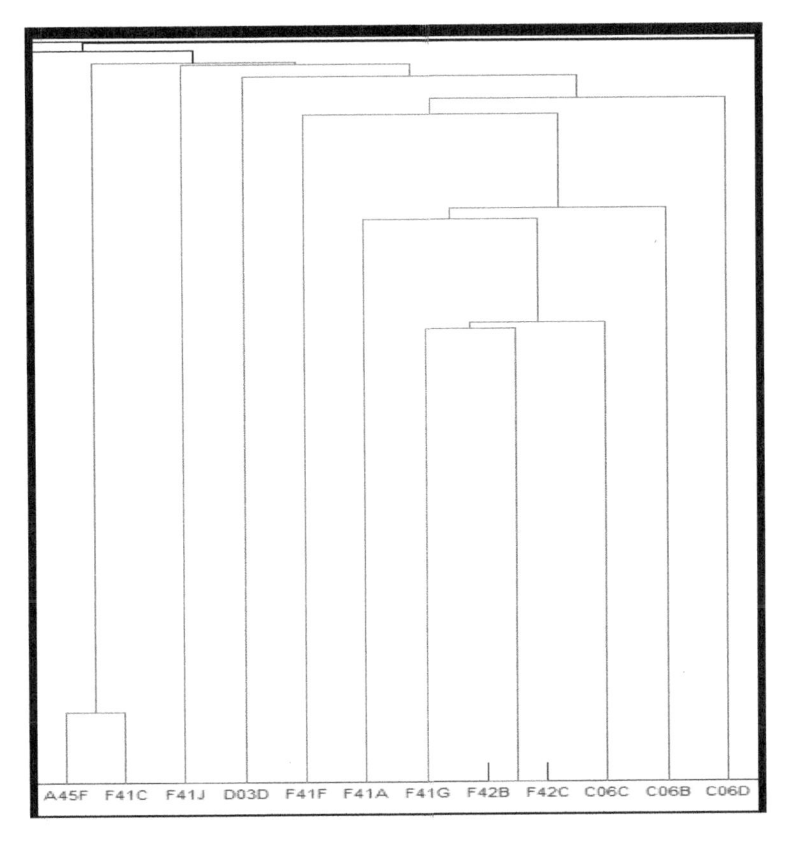


**Figure 3.1.** *Dendrogram portion representing a TKS*

In order to obtain the most relevant results possible, a successive iteration procedure was chosen, which was made possible by the IT resource savings described above. Each iteration corresponds to a cut-off level. The lower the cut-off threshold, the greater the number of technologies included in the analysis. It becomes lower and lower as more technology classes are included in the analysis. Here, the finest cut-off level chosen is 35% of the maximum score and then decreases by 5% increments and then ends, at the smallest level, at a score of 1%. On the one hand, this makes it possible to focus, as a priority, on the strongest synergies within the defence technological knowledge base (12 technologies at the 35% threshold). On the other hand, it allows us to visualize the interdependencies that are grafted onto the former as the number of technologies studied increases. Ultimately, these two elements make it possible to choose the most relevant cut-off level for each TKS. In such an approach, the TKSs whose synergies appear at the highest cut-off thresholds are the most synergetic in the defence knowledge base and therefore the most characteristic of them.

The application of the methodology made it possible to identify 26 TKSs based on the synergies observed within the defence knowledge base between 2010 and 2012. They cover diversified technological fields that go beyond the technological classes belonging to sub-domain devoted to "space and armament" technologies identified by the Observatoire des Sciences et des Techniques (OST).

These 26 TCS are listed in Table 3.1, where they are classified according to the strength of their synergies, i.e. according to the cut-off level at which they appear. The name of each of the TKSs is given for information only, it refers to the most obvious field of application, but all TKSs can, by definition, find applications in different technological systems.

| TKS | Technologies | Technology class number | Number of technologies |
|---|---|---|---|
| 1 | Missiles | F41F F41A F41A F41G F42B F42C C06B C06C C06D | 8 |
| 2 | Armament* | A45F F41C F41J D03D F41F F41F F41A F41G F42B F42C C06B C06C C06D | 12 |
| 3 | Reactors and turbines | F03B F16J F23C F02K F04D F04D F01D F02C F23R F23D F23N F24H F02G | 12 |
| 4 | Defence vehicle | F41H B60D B62D B62D B60G F16F | 5 |
| 5 | Communication | G06Q G06F H04L H04L H04B H04W H04J | 6 |
| 6 | Guidance | H01P H01Q G01S G01C G08G G05D | 6 |
| 7 | Aircraft | B64C B64D | 2 |
| 8 | Remote guidance* | G06Q G06F H04L H04L H04B H04W H04J H04J H03H H01P H01P H01Q G01S G01C | 13 |

| | | | |
|---|---|---|---|
| | | G08G G05D | |
| **9** | Drone* | H03G H03F H04M H03D H04K G05B G06N G06Q G06F H04L H04B H04W H04J H04J H03H H01P H01Q G01S G01S G01C G08G G05D B64C B64D | 25 |
| **10** | Defence Drone* | F41H B60D B62D B60G F16F H03G H03F H03F H04M H03D H04K G05B G06N G06Q G06F H04L H04B H04W H04J H03H H01P H01Q G01S G01C G08G G05D B64C B64D | 30 |
| **11** | Petrochemicals | B01D B01J C07J C07C C10G C10G C10L C07B C01B C01B | 8 |
| **12** | Pneumatic tires | B29B B32B B29B B29C B29D B60C C08K C08L C08F D02G | 10 |
| **13** | Vehicle running gear* | B29K B28B B29B B32B B32B B29C B29C B29D B60C C08K C08K C08L C08F D02G D07B C08C B29L C08J C08G | 16 |
| **14** | Electrical systems | H02J H02M G05F H02P | 4 |
| **15** | Electric motors* | H01M H02J H02J H02M G05F H02P H02K F03D H02H B60L | 9 |
| **16** | Vehicles in general | B60W B60K F16H F16D B60T | 5 |
| **17** | Electric vehicle* | H01M H02J H02J H02M G05F H02P H02K F03D H02D H02H B60L B60W B60K F16H F16D B60T | 14 |
| **18** | Optics, image data processing | G01N G01J H01S G02B G02F H04N G06K G06T | 8 |
| **19** | Electronics | H01L H05K H01R H01R H02G | 4 |
| **20** | Sensors* | G01N G01J H01S G02B G02F H04N G06K G06T G03B H01L H05K H01R H02G | 12 |
| **21** | Biochemistry | C12M C12P C12P C12N C12Q C07H C40B | 6 |
| **22** | Light signals | B60Q F21S F21S F21V F21L F21Y | 5 |
| **23** | Vehicle organization | A47C B60N B60N B60R E06C | 4 |
| **24** | Metalworking | B21B C21D C21D C22C C22F B22F | 5 |
| **25** | Combustion engine | F16N F01M F01L F01L F02D F02D F02B F02M F16K F01N | 8 |
| **26** | Metalworking | B23C B24D B24B B24B B23F | 4 |

**Table 3.1.** *List of TKSs*

There are two categories of TKS. The first is made up of simple TKSs, i.e. independent of any other TKS; the second is made up of complex TKSs, i.e. they are made up of several TKSs grouped together (always according to the strength of synergies). These two levels of TKS make it possible to understand knowledge synergies at different levels and thus highlight that technological groupings identified at different scales can be linked to different technological systems. Complex TKSs are marked with an asterisk in Table 3.1.

TKS consists of an average of 9.2 technologies with just over 6.1 technologies for simple TKS and 16.3 technologies for complex TKS. As expected, the TKSs identified at the highest level of synergy are those with the strongest military specificity (missiles and armaments in general). It is not surprising, however, to find many TKS in more generic areas of knowledge such as electronics, propulsion systems or vehicles. Technologies with a strong link to the aeronautics sector are particularly represented with TKSs 3, 7, 9 and 10. This is logical given the selection of firms and their membership in the A&D sector. However, not limiting ourselves to companies in this sector allows us to highlight greater technological diversity within the defence knowledge base.

For example, TKSs referring to communication systems (TKS5) and guidance systems (TKS6) show that defence manufacturers play a role in the development of these technologies, but the fact that they are mainly synergistic with the aircraft-related TKS (TKS7) shows that this involvement is mainly associated with the development of aircraft distance guidance systems and certainly drones. The synergies between these three TKSs and TKS4, the only TKS whose technological classes explicitly refer to the defence field of application (armor; armored turrets; armored or armed vehicles; means of attack or defence, e.g. camouflage, in general) shows that in these fields of innovation, defence specificity does not constitute the core of the system, but instead, one of its peripheral characteristics. This is consistent with the anticipated duality of these technologies.

The identification of the TKSs, characteristic of defence innovation, initially makes it possible to better understand the synergies that structure its innovation, but it also makes it possible to consider how these synergetic groups can interact with the rest of the technological world. Since the 1980s, the context in which defence innovation has evolved has become increasingly dual (Meunier, 2017), marked by the complexity of equipment (Hobday, 2005), rising costs (Serfati, 2008) and the end of arsenal logic and the privatization of companies (Blanc, 2000). This new environment is pushing companies, while respecting the constraints of national sovereignty, to be more open to technology because of duality and more open with regard to territory because of the internationalization of firms. In this context, it has become more difficult to imagine that military research and development (R&D) develops without taking into account its environment. It is the integration into this environment that offers new opportunities for innovation. This is related to the fact that knowledge production is a cumulative process and a process based on a co-relational structure (Krafft *et al.*, 2011):

– A cumulative process. In other words, the knowledge you are able to produce depends on the knowledge you have acquired in the past.

– A process based on a co-relational structure. The mastery of additional knowledge makes it possible to multiply the combinations of knowledge and thus increase innovation potential accordingly (Fleming and Sorenson, 2001).

In order to integrate with its environment, defence innovation must share a certain amount of knowledge, forming the basis for this integration. The diffusion of knowledge is essential in a technological integration process. However, the dissemination of knowledge depends on the compatibility of the knowledge that the organization masters and the knowledge it desires. This determines the absorption capacities of firms (Cohen and Levinthal, 1990). Therefore, it is important to know the characteristics of defence innovation in order to know, on the one hand, what its absorption capacities are, i.e. the knowledge it is capable of assimilating, and on the other hand, whether the absorption capacities of the rest of the technological domain make it possible to assimilate the knowledge produced in the context of defence innovation. A better understanding of the nature of the knowledge produced within the defence knowledge base therefore provides a better understanding of its integration potential.

Using the typology proposed by Henderson and Clark (1990), it is possible to analyze the production of knowledge within an organization using the notion of knowledge base already mentioned. It makes it possible to highlight the individual knowledge that an organization mobilizes as well as the way in which it organizes it. Then, as Ulrich (1995) did, it is possible to study knowledge blocks and architectural knowledge at the product level. This architecture is understood in this case as the diagram that links these components together. Use of IGT applied to TKS is a new way of identifying knowledge blocks and architectural knowledge by freeing analyses from artifact identification.

Each TKS corresponds to a list of knowledge blocks linked together by an architecture highlighted by technological synergies (citations). It is thus an instrument for identifying the proximity of knowledge that can exist between a TKS and any knowledge base at the level of an organization, an industrial sector or even a territory. These proximities form the basis on which defence innovation can develop its technological and geographical integration.

## 3.6. Conclusion

The EDT and more specifically the IGT offer a toolset to analyze an exchange structure. Applied to the analysis of knowledge flows materialized from patent citations, these tools highlight the relationships of dependence, interdependence and dominance in the process of knowledge production.

This first application of these tools to the defence knowledge base has revealed synergies associated with defence activity within the largest innovative companies in this field. The 26 TKS that were identified in this way provide an image of the specificity of innovation defence among these actors. Another analysis specifically devoted to smaller companies would complete this technological panorama.

TKS is the basic empirical material from which further studies can be conducted. Indeed, studying the TKSs makes it possible to identify, within the sample of companies studied, but above all beyond this predefined scope, which territories, industrial sectors, technologies or companies can contribute to defence innovation. Such studies would allow companies and public authorities to better understand and adapt to the context of dualization and internationalization of the industry.

## 3.7. References


ALIC, J.A., *Beyond Spinoff: Military and Commercial Technologies in a Changing World*, Harvard Business Press, 1992.

AUTANT-BERNARD, C., COWAN, R., MASSARD, N., "Editors' introduction to spatial knowledge networks: Structure, driving forces and innovative performances", *The Annals of Regional Science*, no. 53-2, pp. 315–323, 2014.

BAINEE, J., Conditions d'émergence et de diffusion de l'automobile électrique: une analyse en termes de "bien-système territorialisé", PhD thesis, Université Paris 1 - Panthéon-Sorbonne, 2013.

BIJKER W.E., "How is technology made? That is the question!", *Cambridge Journal of Economics*, vol. 34, no. 1, pp. 63–76, 2010.

BLANC, G., "Dépenses militaires, restructuration de l'industrie d'armement et privatisation de la défense: Analyse comparée France-Etats-Unis, 1994–1999", *Arès*, no. 46, 2000.

BRESNAHAN, T.F., TRAJTENBERG, M., "General purpose technologies 'Engines of growth'?", *Journal of Econometrics*, vol. 65, no. 1, pp. 83–108, 1995.

BUESA, M., "Controlling the international exchanges of armaments and dual - use technologies: The case of Spain", *Defence and Peace Economics*, vol. 12, no. 5, pp. 439–464, 2001.

CARLSSON, B., STANKIEWICZ, R., "On the nature, function and composition of technological systems", *J Evol Econ*, vol. 1, no. 2, pp. 93–118, 1991.

COHEN, W.M., LEVINTHAL, D.A., "Innovation and learning: The two faces of R&D", *The Economic Journal*, pp. 569–596, 1989.

COURNOT, A.A., *De l'origine et des limites de la correspondance entre l'algèbre et la géométrie*, Hachette, Paris, 1847.

DE BANDT, J., "L'émergence du nouveau système technique ou socio-technique", *Revue d'économie industrielle*, vol. 100, no. 1, pp. 9–38, 2002.

FLEMING, L., SORENSON, O., "Technology as a complex adaptive system: Evidence from patent data", *Research Policy*, vol. 30, no. 7, pp. 1019–1039, 2001.

FREEMAN, C., SOETE, L., *The Economics of Industrial Innovation*, Psychology Press, Hove, 1997.

GILLE, B., *Histoire des Techniques: Techniques et civilisation, technique et sciences*, Encyclopédie de la Pléiade, Paris, 1978.

GUMMETT, P., REPPY, J. (eds), *The Relations between Defence and Civil Technologies*, Springer Netherlands, Dordrecht, 1988.

HENDERSON, R.M., CLARK, K.B., "Architectural innovation: The reconfiguration of existing product technologies and the failure of established firms", *Administrative Science Quarterly*, vol. 35, no. 1, p. 9, 1990.

HOBDAY, M., DAVIES, A., PRENCIPE, A., "Systems integration: A core capability of the modern corporation", *ICC*, vol. 14, no. 6, pp. 1109–1143, 2005.

JAFFE, A.B., TRAJTENBERG, M., *Patents, Citations, and Innovations: A Window on the Knowledge Economy*, MIT Press, Cambridge, 2002.

KRAFFT, J., QUATRARO, F., SAVIOTTI, P-P., "The knowledge base evolution in biotechnology: A social network analysis", *Economics of Innovation and New Technology*, vol. 20, no. 5, pp. 445–475, 2011.

LANTNER, R., *Théorie de la dominance économique*, Dunod, Paris, 1974.

LANTNER, R., LEBERT, D., "Dominance et amplification des influences dans les structures linéaires", *Économie appliquée — ISMEA*, vol. 68, no. 3, pp. 143–165, 2015.

LEBERT, D, YOUNSI, H.EL., LEQUEUX, F. *et al.*, "Les échanges industriels entre les pays du bassin méditerranéen: Application d'un nouvel algorithme de clustering sur données de flux", in HAMMOUDA H.B., NASSIM, O., SANDRETTO R. (eds), *Emergence en Méditerranée: attractivité, investissements internationaux et délocalisations*, L'Harmattan, Paris, 2009.

LEBERT, D., YOUNSI, H.EL., "Théorie de la dominance économique: indicateurs structuraux sur les relations interafricaines", *Économie appliquée — ISMEA*, vol. 68, no. 3, pp. 167–186, 2015.

LEYDESDORFF, L., ALKEMADE, F., HEIMERIKS, G. *et al.*, "Patents as instruments for exploring innovation dynamics: Geographic and technological perspectives on 'photovoltaic cells'", *Scientometrics*, 2014.

MERINDOL, V., FORTUNE, E., "Analyse des informations disponibles dans la base Patstat", OST - Observatoire des sciences et des techniques, 2009.

MEUNIER, F.-X., L'innovation Technologique duale: une analyse en terme d'influence et de cohérence, PhD thesis, Université Paris1 Panthéon Sorbonne, 2017.

NESTA, L., SAVIOTTI, P.P., "Coherence of the knowledge base and the firm's innovative performance: Evidence from the U.S. Pharmaceutical Industry", *The Journal of Industrial Economics*, vol. 53, no. 1, pp. 123–142, 2005.

PERROUX, F., "Esquisse d'une théorie de l'économie dominante", *Économie appliquée — ISMEA*, vol. 1, nos 2–3, 1948.

PISCITELLO, L., "Corporate Diversification, Coherence and Firm Innovative Performance", *Revue d'économie industrielle*, vol. 110. pp. 127–148, 2005.

SERFATI, C., "Le rôle de l'innovation de Défense dans le système national d'innovation de la France", *Innovations*, vol. 28, no. 2, p. 61, 2008.

ULRICH, K., "The role of product architecture in the manufacturing firm", *Research Policy*, vol. 24, no. 3, pp. 419–440, 1995.

VERSPAGEN, B., "Structural Change and Technology: A Long View", *Revue économique*, vol. 55, no. 6, p. 1099, 2004.

# 4

# Defence Aerospace Firms: What Are the Technological Coherence of Their R&D?

ABSTRACT. Despite the economic and security challenges related to the French defence technological and industrial base, in particular for defence aeronautics (DA), this DITB has been poorly characterized from the point of view of its technologies (production and use). This chapter aims to fill some of this gap. We focus on the organization of the R&D of DA companies and their production of innovations via a patent approach. We propose a typology of those companies according to their capacity to generate technological synergies in their innovation activity and the trade-offs they make between exploration innovations and exploitation innovations.

## 4.1. Introduction

Defence aeronautical firms are intensively involved in research and development (R&D). They seek to push back the technological boundaries. In this respect, they are in line with the defence policy's desire to satisfy strategic autonomy for national armies [MIN 13]. Armament programs, which are key elements of the defence industry, are largely based on technology control considerations [SER 14]. R&D also responds to public innovation policy, which gives defence firms a role as technological coaches [SER 08]. At the same time, it allows them to gain competitiveness in various markets, especially export markets [FON 14].

Chapter written by Cécile FAUCONNET, Didier LEBERT, Célia ZYLA and Sylvain MOURA.

Despite the economic and security challenges faced by the French defence technological and industrial base (DTIB), and in particular by the defence aeronautics (DA) sector, the latter has been poorly characterized in terms of its technologies (production and use). The focus was mainly on the conditions for financing defence firms [BEL 08, OUD 15]. Only one case study on the Thales group [AYE 08] takes a technological approach, as the focus is mainly on the management of the resulting property rights.

This chapter aims to analyze defence aeronautical firms in terms of their technologies. We will focus on the organization of their R&D and the production of innovations through a patent approach. Two topics are addressed in more detail in order to identify specificities these firms have. First, what are the relationships between their technological duality (their ability to produce military and civil technological blocks) and their productive duality (their ability to generate military and civil turnover)? Second, what links do they form between their technological diversity (the number of technological knowledge blocks they produce) and the nature of their R&D (oriented more towards the exploration of original technological knowledge architectures or towards the exploitation of controlled architectures)?

This study focuses on the technological organization required for firms' innovations, despite their diversification in terms of size, R&D structure, strategy, etc. We introduce a multidimensional measurement of their technological coherence [NES 05] and deduce from it a typology for DTIB firms in the DA domain, according to their ability to generate synergies in their innovation activity, the trade-offs they make between exploration and exploitation innovations, the technologies they handle (in number and scope), their size, their defence dependence rate, etc. Defence aeronautical firms are identified through their APE business identifier code[1] and their participation in military aeronautical programs.

In this chapter, section 2 sets out the working hypotheses in relation to the two topics described above: the relationship between technological duality and productive duality, on the one hand, and between technological diversity and the management of the exploration/exploitation dilemma, on the other hand. Section 3 provides an update on the measurement of

1 2630Z (Manufacture of communication equipment), 2651A (Manufacture of navigational aids equipment) and 3030Z (Manufacture of aircraft and spacecraft).

technological coherence and the exploration ratio. Section 4 presents a sample of firms from the DTIB and the associated data. Finally, section 5 is devoted to the main results of the study.

## 4.2. Assumptions on the relatedness and technological coherence of DA firms

The relationship between the duality of activities and technological duality can be addressed in an analytical framework based, in industrial economics and strategic management, on the link between diversification or refocusing on the economic and financial performance of firms (even though we do not consider in this chapter to associate this relationship with performance).

In line with Ansoff's work [ANS 65], the proposal that conglomerate diversification ensures that the firm minimizes its systematic risks is questioned in light of the potentially negative impact that this strategy may have on the firm's performance. Ansoff highlights the need to combine the practice of diversification with the given state of their business portfolio in order to maximize productive and commercial synergies. There are therefore two diversification strategies with a differentiated theoretical effect on the firm's performance: conglomerate diversification, or "unrelated", versus "related" diversification. As shown in [ELY 13], the related diversification strategy (or refocusing strategy) may not be accompanied by an increase (or decrease) in the number of activities in the firm's portfolio. These strategies thus have two dimensions that can be dissociated: the evolution of the number of activities – i.e. quantitative diversification (or refocusing) – and the evolution of the intensity of the links between activities – i.e. qualitative diversification (or refocusing).

Measuring the intensity of the link between the company's productive activities has given rise to abundant literature. The most common way to do this is to calculate distances between activities in the same portfolio according to whether they belong to more or less identical fields in an external activity classification. For example, [PAL 85] uses the American Standard Industrial Classification (SIC) of activities by considering strong links to be those between activities that belong to three-digit identical levels of aggregation, weak links for those between activities that belong to two-digit identical levels of aggregation and nil for a higher degree of aggregation. Entropy indices specific to the related and unrelated portfolio activities are then calculated (with the turnover of the different segments used to quantify each of the activities), and the theoretical relationship tested is that between the intensity of the links and the value of the firm's ROA (return on assets) or ROI (return on investment).

The use of entropy indices to understand these linkages is still widespread today [ELY 14]. Other approaches have also been proposed. For example, the studies conducted by Rumelt [RUM 74, RUM 82] and Montgomery [MON 79, MON 94] address linkages through strategic categories that combine the degree of specialization of the firm (in terms of turnover of its main activity) with the degree of connectivity of the activities (according to the identity of the resources used to operate them). With this literature, we return to Ansoff's primary concerns who considered that business development strategies were mainly guided by this degree of resource connectivity. This will be shown empirically [CHR 81] for diversification and [KLE 09] for refocusing (firms withdraw as a priority from activities that use few common resources with those employed in their core business).

The link between activities is therefore conceived here at the resource level. Multiple entries have been explored, and [TAN 05] try to provide an overview of them: human resources, technology, etc. Concerning the "technology" resource, three inputs are preferred. [ROB 95] first of all, to rely on Scherer's [SCH 82] technology flow matrices constructed from patent data to calculate the intensity of linkages between productive activities. [RON 05] for their part, following [FAN 00], produce technology linkage indices from data obtained at the macro level – from the national accounts (inter-industry trade tables). The extensions of the coherence approach [PIS 00, PIS 04] (see section 3) also mobilize patent data to define the firm's innovation activity. In all three cases, the "technology" linkage fully determines the "activity" linkage, so that a direct relationship is established between the former and the firm's performance.

For Robins and Wiersema, Piscitello, or Rondi and Vannoni, they take into account only the qualitative dimension of their development strategy and therefore exclude diversity of activities themselves. At best, this diversity is used as a relatedness indicator to normalize its value and make comparisons between companies with different productive diversity.

The distinction between technological duality and activity duality requires the reintroduction of the quantitative dimension.

As anticipated from the above, this distinction can be made in two different ways: by counting activities (diversity *sensu stricto*) or by reasoning in terms of turnover shares (specialization in the Rumelt sense). It is this second approach that we will adopt here by assuming that a defence firm with a "diversified" activity is one that has a specialization ratio in defence activities below a certain threshold. It is then dual in terms of its activities because a significant part of its turnover comes from civil markets.

We use the [ROB 95] method for technological duality: we favor an approach that focuses on the destination in terms of technology activities rather than on the organization of R&D processes (more or less "coherent") as conceived by Piscitello. From our point of view, the organizational coherence of firms' R&D processes refers to issues that relate to strategic choices of a different nature than those of duality. We will come back to this later.

Technological duality refers to the number of technological blocks mastered by the firm that are intended for military use to all the technological blocks dealt with by the firm. A firm with a defence activity is technologically dual if the technologies it masters include a significant proportion of non-defence technologies.

Cutting the exclusive link between the resource and the productive and commercial activity does not mean that all "dual/non-dual" combinations can be found in the same proportions. We thus assume that technological duality is positively associated with the duality of activities. This hypothesis results from reading the references mentioned above, which strictly link the linkages at these two levels, until they are muddled.

[PIS 00, PIS 04, NES 05] and [KRA 11] come from the same initial literature, with the same aims (to identify the links between the effort and the nature of the firm's innovation and its performance), but focus on the representation of the organization of a firm's R&D. For this purpose, they draw their inspiration from the model of the coherence of large companies developed by [TEE 94].

[ELY 15] show that it is possible to extend this representation to consider the concrete organization of firms' R&D in terms of exploration/exploitation dilemmas [MAR 91]. It then becomes possible to return to the observations made by [QUI 08] for whom technological diversity has a greater effect on the firm's exploratory capacity than on its exploitation capacity. What about firms with a defence activity?

In the following section, we present how to calculate a firm's exploration capacity by placing ourselves in line with [TEE 94] and the work that this article has inspired. Like in [QUI 08] and [BOT 10], the technological diversity of the firm corresponds to the number of technological blocks of a different nature that it is able to produce. With an inherent bias in the strict approach to coherence, found in the latter authors, means that it cannot be applied immediately: the positive link between diversity and coherence that they empirically highlight is based on a purely theoretical organization of the firm's R&D, which assumes that all possible technological linkages – between each technological block produced – are implemented – "exploited". The gap between this maximum exploitation situation and the firm's actual R&D situation represents the degree of exploration in this organization. The hypothesis we are posing – based on [QUI 08] – is that the diversity of the technological blocks produced by the firm is positively associated with this exploration capacity, which can be considered to be a gap between two coherences, one theoretical and the other concrete.

## 4.3. Measuring technological coherence

The coherence measures introduced by [TEE 94] are closely linked to the structural/reticular way of representing firms' knowledge bases [SAV 09]. However, they cannot be perfectly so, because two firms handling the same technologies but linking them in different ways will have the same overall coherence and neighborhood coherence scores (see below). Therefore, we propose to complete the study of the coherence of firms' knowledge bases by articulating:

– global coherence, where all possible linkages between the firms' technologies are taken into account. This coherence is measured by the weighted average relatedness (WAR) indicator of [TEE 94] adapted to the context of technological coherence by [NES 05];

– neighborhood, or "core" coherence, where only the strongest technological associations are taken into account so as to form a "Maximum Spanning Tree". This index corresponds to weighted average relatedness of neighbors (WARN) [TEE 94];

– the coherence of concrete inter-technology relationships, where only the truly identified links between technologies within the firm's knowledge base are taken into account [NAS 13]. We call this indicator KBC (Figure 4.2).

It is through patent data that we represent the knowledge bases of firms in a reticular way. Let $T$ be the total number of different technologies in the patent database, $p_k$ the format vector, $T \times 1$ indicating for each of its arguments the total number of occurrences of $T$ technologies for the firm $k$, $p_{ki}$ this total number for the technology $i$ and $t_k$ the technological scope of the firm (the number of different technologies it produces, i.e. the number of non-zero arguments of $p_k$); $P_k \equiv \sum_i p_{ki}$ corresponds to all technological occurrences in patents held by the firm (Figure 4.1).

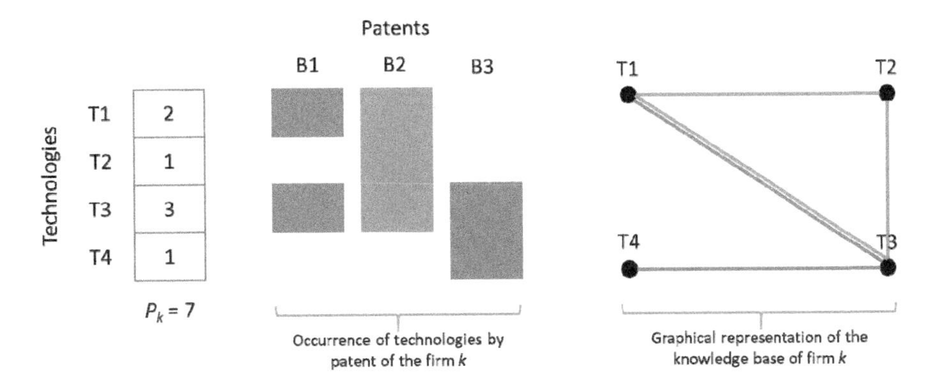


**Figure 4.1.** *Data structure for measuring the technological coherence of a firm. For a color version of the figures in this chapter see www.iste.co.uk/barbaroux/technology.zip*

Let us call $C_k$ the coherence index for the firm $k$; this index can take three distinct forms depending on the content of $M^k$, the Boolean matrix of format $T \times T$, which summarizes the technological linkages according to each of the three configurations of coherence for the firm $k$. Therefore:

$$C_k = \frac{1}{P_k} \sum_i p_{ki} \left( \frac{\sum_{j \neq i} \tau_{ij} M_{ij}^k p_{kj}}{\sum_{j \neq i} M_{ij}^k p_{kj}} \right) \qquad [4.1]$$

Each firm is thus characterized by three absolute coherence indicators (according to the shape of the matrix $M$), and two in relative value (exploration and exploitation rates, see Figure 4.2):

– for *WAR*, the adjacent matrix $M$ is only composed of 1 to integrate all possible linkages between the technologies dealt with by $k$;

– for *WARN*, the matrix $M$ integrates the strength of technological linkages into the firm's innovation production space. While *WAR* combines the firm's technologies by $t_k(t_k - 1)/2$ links, here we associate all of them in $t_k - 1$ links only (i.e. in a tree) by retaining only the association – $\tau_{ij}$ being the value of this strength [BOT 10] – which gives the total weighting of the strongest tree. In other words, the definition of $M$ is based on a weighted maximum spanning tree procedure and $M_{ij}^k = 1$ if $ij$ belongs to the maximum spanning tree, 0 otherwise;

– for *KBC*, the matrix $M$ summarizes the concretely identified linkages between technologies within the firm's knowledge base, with $M_{ij}^k = 1$ if the linkage $ij$ actually belongs to the knowledge base of $k$, 0 otherwise. This is the indicator of "technological complementarity" in [NAS 13].

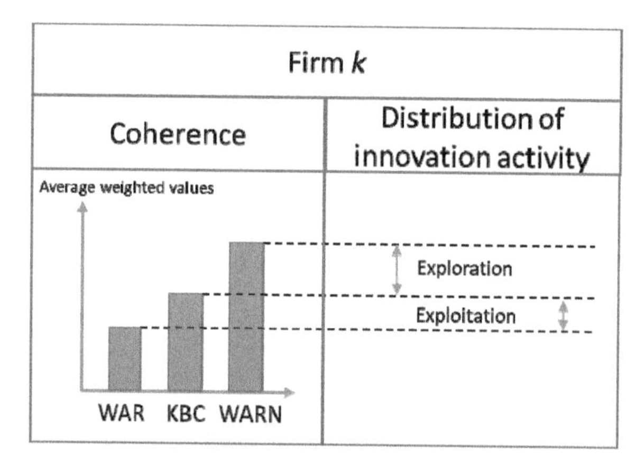


**Figure 4.2.** *Exploration and exploitation in a firm's knowledge base*

[TEE 94] show that *WAR* is necessarily less than or equal to *WARN*. The general case shows that KBC falls within this range (Figure 4.2). However, the $WAR \leq KBC \leq WARN$ sequence is not always verified, especially for firms that deal with disconnected knowledge blocks (without any direct and indirect relationship between them).

Based on the idea of [ELY 15], the firm's strategic decision regarding the organization of its knowledge base involves positioning itself in relation to the "core" of the technologies it combines, i.e. in relation to *WARN*. The firm can indeed make two choices in relation to this core:

– "explore unusual technological associations". These are not the most frequent ones found in the entire patent sample. In this sense, these linkages represent a "differentiation" factor for the firm;

– "not to exploit obvious technological associations". They are often found in the patent sample and can be expected to contribute relatively easily to the creation of innovation locally. Not exploiting them is in this sense is a "distinction" factor for the firm.

Distinction and differentiation reflect the firm's exploration activity. The exploitation activity – measured by the difference between *KBC* and *WAR* – reveals the weighting of the "core" in the concrete organization of its knowledge base.

The two relative indicators are therefore:

– the exploration rate: $EXPLOR = (WARN - KBC) / (WARN - WAR)$;

– the exploitation rate: $EXPLOIT = 1 - EXPLOR = (KBC - WAR) / (WARN - WAR)$.

The sum of their value is equal to 1.

## 4.4. The data: scope and content

The starting population is the defence industrial and technological base [MIN 17]. It spans the 2012–2014 period. The analysis concerns a sub-population of this DTIB, i.e. 81 legal units (identified by their SIREN number). They were selected in two stages. The purpose of this selection is to focus the analysis on the units involved in aeronautics, i.e. the construction of aircraft and their movement in the Earth's atmosphere.

First, the legal units belonging to the manufacturing industry that supply the French Ministry of Defence in 2016 for military aeronautical programs have been isolated. Military aeronautics includes helicopter, drones, combat and transport aircraft programs, as well as associated electronics. It includes design (including R&D), production and maintenance (without dismantling equipment). Missiles (conventional and ballistic) and space are excluded from the field.

In the second phase, further filtering was carried out to retain suppliers who belonged to activity sectors with the APE code 2630Z (manufacture of communication equipment), 2651A (manufacture of navigational aids equipment) and 3030Z (aircraft and space construction). These are the typical sectors of aeronautics, which is reflected by their contribution: in 2016, they represented 91% of the French Ministry of Defence's financial support to the sectors identified in the first selection stage (via military aeronautics programs).

We retain two variables from the DTIB. First of all, the average turnover of firms between 2012 and 2014 (rated CA) to assess their size, then a binary variable that reflects the commercial independence of the firm from its military activity (rated INDEP). The value of 0 is assigned to INDEP if the share of military turnover in the firm's total turnover is greater than 20%, otherwise its value is set at 1 (the INDEP variable can only be treated discretely due to data confidentiality).

We use patents as a proxy for technological innovation. According to [GAR 79], they are a good approximates. Patents are intellectual property rights that provide protection in one or more countries. Inventions are often the subject of several patent applications to various national or regional offices or to the World Intellectual Property Organization (WIPO), the global forum for intellectual property services, policies, information and cooperation. Therefore, analysis in terms of patent *families* is relevant when looking at innovation. Indeed, a patent family makes it possible to capture an effective innovation and not simply the reproduction of an innovation in multiple national or regional contexts. Questel, the publisher of the ORBIT database we use, has developed a family perimeter that combines the EPO's strict family rule[2] with additional rules that take into account links with the parent application Europe and/or PCT[3], as well as links between provisional

2 The European Patent Office (EPO) defines a patent family as including all documents with the same priority or combination of priorities.

3 "The *Patent Cooperation Treaty (PCT) helps applicants obtain patent protection at the international level, assists patent offices in their patent granting decisions, and facilitates public access to a wealth of technical information about these inventions. By filing a single international patent application under the PCT, applicants can apply for protection of an invention simultaneously in 148 countries around the world*" (http://www.wipo.int/pct/fr/, accessed on 03/08/2016).

US applications and published US applications. This scope also takes into account the different definitions of invention according to the offices.

To identify the technologies contained in patents, we use the International Patent Classification published by WIPO. WIPO divides the technology into eight main sections (see Figure 4.3) with about 70,000 subdivisions at the most disaggregated level.

Over the period 2012–2014, we identified via ORBIT-Questel 3,558 patent families (first filing dates), composed of 10,913 patents concerning 35 of the 81 firms in our analysis. These firms produce (families of) patents in all the aggregate technology fields defined by WIPO (main sections; see Figure 4.3), with a dominance in the main technology of industrial techniques (35%), closely followed by mechanical, lighting, heating and weapons technologies (31%). We also see a strong presence of families associated with physics (19%) and electricity (11%).

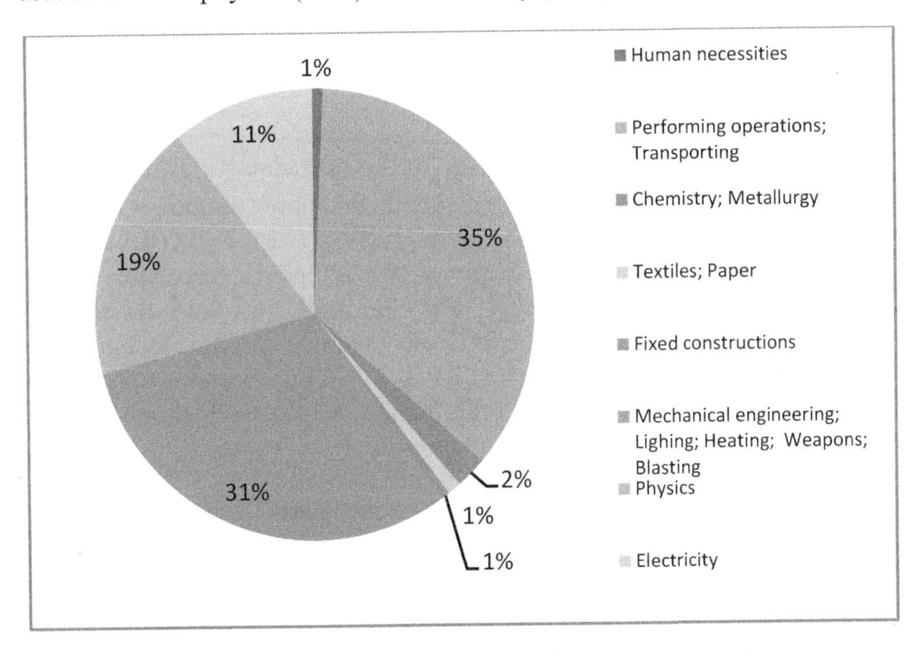


**Figure 4.3.** *Distribution of patent families in the sample by major technology sections (main technological area)*

Measuring the technological coherence of firms requires building a global technological landscape using another data set. Index values $\tau_{ij}$ are calculated using the [NES 05] method and data from the PATSTAT database (autumn 2014 version) published by the European Patent Office (EPO). PATSTAT is a global database containing patent applications from more than 80 countries. It contains more than 80 million records and provides information such as inventors and owners, technology classes, titles and abstracts, publication offices and citations.

We extract all patents applied for between 2010 and 2012 from the European (EPO), American (USPTO) and EU15 Member States patent offices. We group these patents into International Patent Documentation (INPADOC) extended families (patents that have at least one priority in common, directly or indirectly). The 2010–2012 results were chosen as a result of the density of the data obtained in order to produce technological landscapes rich enough to have the most accurate coherence indicators possible. There were more than 327,000 families in 2010, 336,000 in 2011 and 243,000 in 2012, for a total of about 907,000 different families.

The index $\tau_{ij}$ describes the intensity of the relatedness existing between the technologies $i$ and $j$. The higher its value, the more common it is to find these two technologies within the same family. Each INPADOC family contains all the technologies it requests and is updated according to the WIPO nomenclature. We adopt the four-digit breakdown of this nomenclature, the level most commonly adopted in the literature on innovation economics, for example [KRA 11]. This breakdown identifies 643 different technologies, and each family is a combination of these technologies in the format shown in Figure 4.1.

Of these 643 technology entries, 16 are part of the "Space and Armaments" technology sub-group according to [OST 10]. It is these 16 entries that we use to list the "defence" technologies within the families extracted from ORBIT for our sample of firms (Table 4.1). The share of "defence" technologies in the total number of technologies produced by DA firms will make it possible to calculate a technological duality indicator.

| Code | Title |
|---|---|
| **B63G** | Offensive or defensive arrangements on vessels, mine-laying, ... |
| **B64G** | Cosmonautics, vehicles or equipment therefor (apparatus for, or methods ... |
| **C06B** | Explosive or thermic compositions (blasting f42d), manufacture thereof, ... |
| **C06C** | Detonating or priming devices, fuses, chemical lighters, pyrophoric ... |
| **C06D** | Means for generating smoke or mist, gas-attack compositions, generation ... |
| **C06E** | Matches, manufacture of matches |
| **F41A** | Functional features or details common to both small arms and ordnance, ... |
| **F41B** | Weapons for projecting missiles without use of explosive or combustible ... |
| **F41C** | Small arms, e.g. Pistols, rifles (functional features or details common ... |
| **F41F** | Apparatus for launching projectiles or missiles from barrels, e.g. ... |
| **F41G** | Weapon sights, aiming (optical aspects thereof g02b) |
| **F41H** | Armor, armored turrets, armored or armed vehicles, means of attack or ... |
| **F41J** | Targets, target ranges, bullet catchers |
| **F42B** | Explosive charges, e.g. For blasting, fireworks, ammunition (explosive ... |
| **F42C** | Ammunition fuses (blasting cartridge initiators f42b0003100000, chemical ... |
| **F42D** | Blasting (fuses, e.g. Fuse cords, c06c0005000000, blasting cartridges ... |

**Table 4.1.** *The technological subclasses space and armament [OST 10]*

Table 4.2 presents descriptive statistics on the turnover and number of patent families filed for all the firms studied, as well as a breakdown of the latter according to their level of defence independence (high or low). This table shows a disparity between firms in terms of turnover and a relative homogeneity in terms of patent family filing. If we compare highly independent defence firms with weakly independent firms, we find that the former have a higher average turnover than the latter. On the contrary, firms that are weakly independent of defence are less dispersed in terms of the number of patent families filed than the other firms studied.

| | Turnover (in millions) | | | | | | Patent families | | | | |
|---|---|---|---|---|---|---|---|---|---|---|---|
| | N | Average | Standard deviation | Median | Min | Max | Average | Standard deviation | Median | Min | Max |
| **All firms** | 81 | 525.3 | 1 326.0 | 34.0 | 0.4 | 8 703.8 | 52.4 | 184.3 | 0 | 0 | 1 221 |
| **Firms with low independence** | 39 | 483.2 | 939.2 | 34.4 | 1.6 | 3 500.7 | 51.3 | 163.0 | 0 | 0 | 882 |
| **Strong independence firms** | 42 | 564.3 | 1 615.6 | 29.2 | 0.4 | 8 703.8 | 53.5 | 204.2 | 0 | 0 | 1 221 |

**Table 4.2.** *Descriptive statistics of DA firms*

Table 4.3 presents, for both types of firms, the equality tests of the averages on the capacity to innovate, i.e. the number of patent families filed, and on the size of the firm, i.e. turnover. This table only partially confirms the findings made previously on the basis of descriptive statistics. Indeed, these tests show that firms file on average the same number of patent families but that the disparity in filings is much greater for firms that are highly independent of defence. With regard to turnover, the results highlight a homogeneity between the two types of firms considered.

Table 4.4 presents the average equality tests on the size of the firm, i.e. turnover, and the dual nature of the firms' activities, i.e. independence in defence, between firms filing patents and others. These tests do not show any difference in average in the defence independence between innovative firms and others. On the contrary, we note that innovative firms have a higher average turnover than firms that do not file patents.

| Variable tested: publication of patent families | | | | | | |
|---|---|---|---|---|---|---|
| **Independence rate** | *N* | *Average* | *Standard deviation* | *Pte error* | *Min* | *Max* |
| **0** | 39 | 51.33 | 163 | 26.1 | 0 | 882 |
| **1** | 42 | 53.54 | 204 | 31.51 | 0 | 1 221 |
| | | Average equality | P-value | Equality variances | | |
| | | yes | 0.96 | no | | |

| Variable tested: turnover (in millions) | | | | | | |
|---|---|---|---|---|---|---|
| **Independence rate** | *N* | *Average* | *Standard deviation* | *Pte error* | *Min* | *Max* |
| **0** | 39 | 483.3 | 939.2 | 150.4 | 1.6 | 3 500.8 |
| **1** | 42 | 564.4 | 1 615.7 | 249.3 | 0.4 | 8 703.9 |
| | | Average equality | P-value | Equality variances | | |
| | | yes | 0.78 | yes | | |

**Table 4.3.** *Equality tests on patents and turnover of DA firms according to defence independence*

More than one in two firms (46 out of 81) do not file patents in this high-tech sector. At least three reasons can explain this phenomenon:

– legally, firms manufacturing military equipment are required to declare to the Directorate General of Armaments the patent applications they file (Article L612-8 of the Intellectual Property Code). The management shall decide on the sensitive nature of the invention and may require non-disclosure of the patent. In this case, patents are held incommunicado by the defence authorities for reasons of sovereignty;

– some manufacturers may prefer secrecy as a means of protecting their inventions;

– some firms in the sector, subcontractors or operational maintenance firms, do not produce patentable technological innovations.

| Variable tested: defence independence | | | | | | |
|---|---|---|---|---|---|---|
| **Patent family filing** | *N* | *Average* | *Standard deviation* | *Pte error* | *Min* | *Max* |
| **0** | 46 | 0.57 | 0.50 | 0.07 | 0 | 1 |
| **1** | 35 | 0.46 | 0.50 | 0.08 | 0 | 1 |
| | | *Average equality* | *p-value* | *Equal variances* | | |
| | | yes | 0.34 | no | | |

| Variable tested: turnover (in millions) | | | | | | |
|---|---|---|---|---|---|---|
| **Patent family filing** | *N* | *Average* | *Standard deviation* | *Pte error* | *Min* | *Max* |
| **0** | 46 | 123.7 | 2 690.4 | 39.6 | 0.4 | 1 375.9 |
| **1** | 35 | 1 053.1 | 1 880.7 | 317.8 | 1.8 | 8 703.9 |
| | | *Average equality* | *p-value* | *Equal variances* | | |
| | | no | 0.0014 | yes | | |

**Table 4.4.** *Equality tests on defence independence and turnover of DA firms according to patent family filing*

## 4.5. Main results

In summary, the relationships and variables that define the profiles of DA firms in our study are:

– relationship 1 (R1): technological duality (IPCNONDEF) and productive duality (INDEP). IPCNONDEF is a ratio of the number of non-defence technologies developed by the firm according to the nomenclature of [OST 10] to its total number of technologies developed. INDEP is a Boolean variable indicating the weighting of the firm's non-defence turnover. If at most 20% of its turnover (on average over 2012–2014) comes from the military sphere, the INDEP variable takes the value 1, otherwise it takes the value 0. By hypothesis, these two variables are supposed to act in the same way to define the profile of DA firms (see section 1);

– relationship 2 (R2): technological diversity (DIV) and exploration ratio (EXPLOR). DIV indicates the extent of the firm's technological mastery – the number of different items in the four-digit WIPO nomenclature of patents it produces over the period 2012–2014. EXPLOR is a coherence variable that reflects the importance of the firm's strategic technological differentiation and distinction behaviors. By hypothesis, DIV and EXPLOR are supposed to act in the same way to define the profile of AD firms [QUI 08] if we assume that these firms behave like the others;

– the variables PATENT (number of patents filed in 2012–2014), IPC (number of technologies developed during the same period) and CA (average turnover over 2012–2014) are taken into account to control these relationships. PATENT and IPC are selected because, together, they provide information on the technological scope of the patents filed by the firm. CA is one of the firm's key non-technological characteristics: its size.

Figure 4.4 shows the results of a dual data analysis: factor analysis and hierarchical bottom-up classification. Each firm is represented by a vector comprising seven values, for each of the variables described above. First, a factor analysis projects these vectors into a plan, and the contribution of each variable to the two main factors is indicated. Then, a hierarchical cluster analysis using weighted average distances identifies five business classes (shown in color in Figure 4.4).

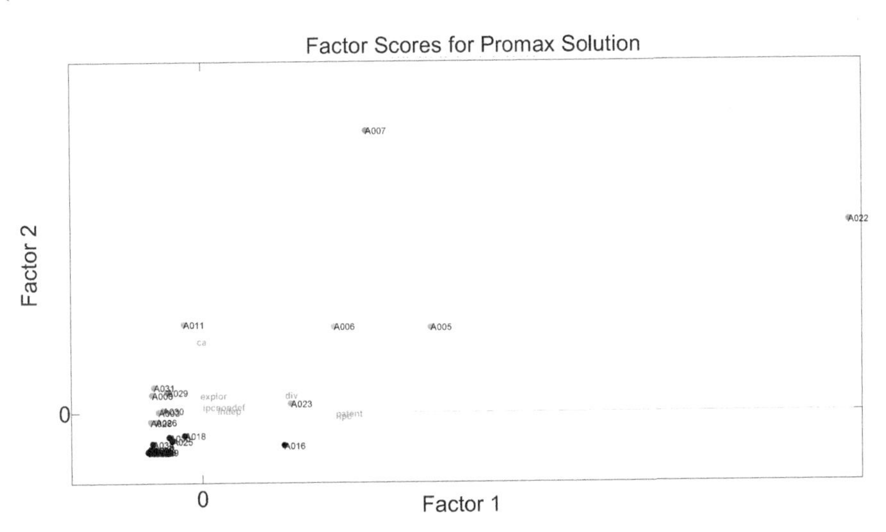


**Figure 4.4.** *Factor analysis on the technological coherence of defence aeronautical firms*

The technological control variables (PATENT, IPC) contribute strongly to defining the first factor, while the non-technological control variable (CA) actively participates in the second. For the R2 relationship, the variable DIV is positively associated with both factors, whereas the variable EXPLOR is positively associated with only the second. Technological diversity and exploration ratio (Figure 4.5 shows the shape of this relationship in our sample) are therefore partly in line with the technological scope and size of the firm. On the contrary, the R1 relationship is not true: IPCNONDEF contributes to the second factor (and therefore associates itself with size), while INDEP does not contribute to any factor.

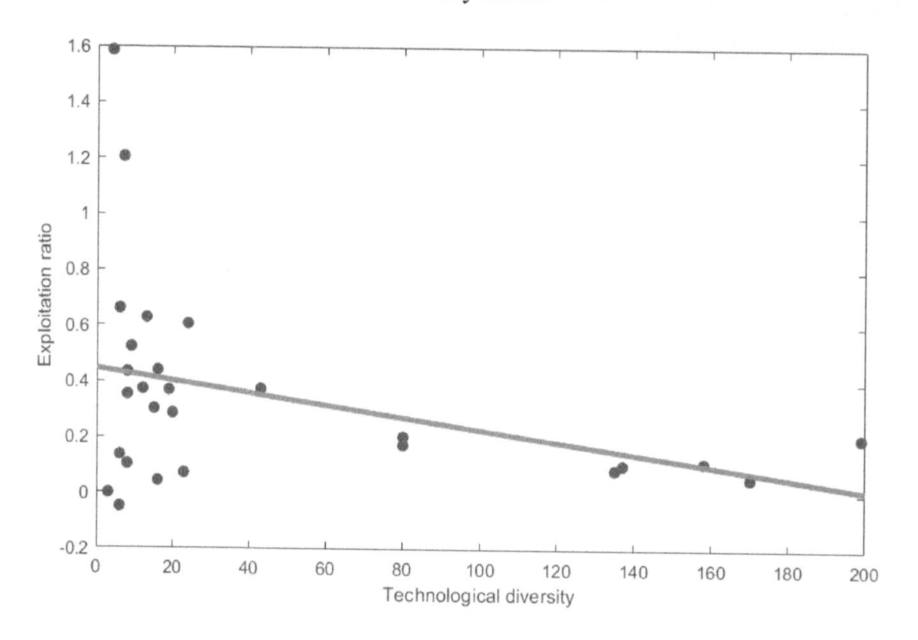


**Figure 4.5.** *Technological diversity and exploitation ratio*

The R1 relationship between the two dualities (technological and productive) is not established for DA firms. This may be due in part to the fact that our duality indicators can be discussed, both in their forms and in the thresholds we have chosen. Concerning their forms, we have justified in section 2 an indicator based on strictly "defence" technologies of the nomenclature maintained by [OST 10]. This technological base (Table 4.1) is probably too narrow, making the IPCNONDEF ratio too high on average.

For the INDEP indicator, the choice to use a Boolean variable is questionable: if the threshold has a maximum of 20% of "military" turnover, assigning a value of 1 to this variable is probably reasonable, using a continuous variable rather than a binary one might have led to richer results.

In the field of defence aeronautics, firms with a very high diversity coexist with *pure players* with narrower markets and more specific productions. For the R2 relationship, the growth in the exploration rate with technological diversity echoes the [QUI 08] approach on several levels. They argue that while technological diversity has a greater effect on the firm's exploratory capacity than on its exploitation capacity; a firm with a diversified technological base also has a higher innovative potential. In addition, this diversity can reduce rigidities and path dependencies by accelerating the firm's "invention rate", especially when it deviates from its routine activities. This opens up new perspectives for the specific study of firms' innovation profiles, particularly in the DA sector.

The most direct of these perspectives is, in our opinion, derived from the work of [ZYL 17]. She starts from the approach of Quintana-Garcia and Benavides-Velasco by developing a representation in which the ability of a firm to deviate from its routines is translated into a "potential for distortion" revealed by the gap between the firm's "technological cores", on the one hand, when this core is defined based on its sole core activity, and on the other hand, when it is defined based on the global technological landscape. By adopting the empirical approach to the firm's technological coherence described in section 3, this amounts to producing indicators $C_k$ when the patent base for measuring indices $\tau_{ij}$ focuses initially only on firms in the company's main sector (e.g. DA) and then on all firms regardless of their main sector. When a firm develops a technological link belonging to the global core without being from the sectoral core, it innovates beyond its core technological competence, and acquires the ability to combine cores in order to produce breakthrough innovations for its core sector. Zyla shows empirically – especially for firms in the aeronautics sector – that the potential for distortion is positively linked to the technological diversity of the firm. This result gives substance to the proposition that there is a positive relationship between technological diversity and invention rates by assigning it a specific empirical content.

## 4.6. Conclusion

This study of the R&D of the DA firms of the French DTIB has highlighted two major elements. On the one hand, the study does not validate the hypothesis of a positive relationship between technological duality and productive duality in this sector. It highlights the weakness of the overly restrictive classification of space and weapons technologies, which does not take into account, for example, all radar technologies (three-digit WIPO codes G01 and G09) that are essential for the DA activity. On the other hand, our second line of research concerning R&D strategies, the relationship between technological diversity and the exploration ratio, is more promising. Indeed, our analysis in terms of technological *relatedness* shows a positive relationship between the number of technologies mastered by firms and the coherence reflected by the "exploration of strange technological associations" (differentiation) and "absence of exploitation of obvious technological associations" (distinction). In the future, it would be interesting to analyze the profiles of DA firms from an international comparative perspective in a context marked by recent developments in this market: the emergence of new competitors, particularly Chinese (Commercial Aircraft Corporation of China), Airbus' commercial repositioning around civil aircraft, etc.

## 4.7. References

[ANS 65] ANSOFF H.I., *Stratégie du développement de l'entreprise*, Edition Hommes et Techniques (French translation, 1981), Paris, 1965.

[AYE 08] AYERBE C., LAZARIC N., CALLOIS M. *et al.*, "Nouveaux enjeux d'organisation de la propriété intellectuelle dans les industries complexes: une discussion à partir du cas de Thales", *Revue d'Economie Industrielle*, no. 137, pp. 9–42, 2008.

[BEL 08] BELIN J., GUILLE M., "R&D et innovation: quel financement pour les entreprises défense ?", *Innovations*, no. 28, pp. 33–59, 2008.

[BOT 10] BOTTAZZI G., PIRINO D., "Measuring industry relatedness corporate coherence", *LEM Working Paper Series*, 2010.

[CHR 81] CHRISTENSEN H.K., MONTGOMERY C.A., "Corporate economic performance: diversification strategy versus market structure", *Strategic Management Journal*, vol. 2, pp. 327–343, 1981.

[ELY 13] EL YOUNSI H., *Les stratégies de recentrage des entreprises: Europe, États-Unis, 1990–2008*, Presses Académiques Francophones, Paris, 2013.

[ELY 14] EL YOUNSI H., LEBERT D., "Un retour au ratio de spécialisation pour interpréter les stratégies de développement des entreprises non financières: Europe et États-Unis, 1992–2007", *Economie Appliquée*, vol. 67, no. 1, pp. 5–36, 2014.

[ELY 15] EL YOUNSI H., LEBERT D., MEUNIER F.-X. *et al.*, "Exploration, exploitation et cohérence technologique", *Economie Appliquée*, vol. 68, no. 3, pp. 187–204, 2015.

[FAN 00] FAN J., LANG L., "The measurement of relatedness: An application to corporate diversification", *The Journal of Business*, vol. 73, pp. 629–660, 2000.

[FON 14] FONFRÍA A., DUCH-BROWN N., "Explaining export performance in the Spanish defence industry", *Defence and Peace Economics*, vol. 25, pp. 51–67, 2014.

[GAR 79] GARFIELD E., "Is citation analysis a legitimate evaluation tool?", *Scientometrics*, vol. 1, pp. 359–375, 1979.

[KLE 09] KLEIN P.G., KLEIN S.K., LIEN L.B., "Are divestitures predictable? A duration analysis", *Working Paper*, Contracting and Organizations Research Institute, University of Missouri, 2009.

[KRA 11] KRAFFT J., QUATRARO F., SAVIOTTI P.P., "The knowledge base evolution in biotechnology: A social network analysis", *Economics of Innovation and New Technology*, vol. 20, pp. 445–475, 2011.

[MAR 91] MARCH J.G., "Exploration and exploitation in organizational learning", *Organization Science*, vol. 2, Special Issue: Organizational Learning: Papers in Honor of (and by) James G. March, pp. 71–87, 1991.

[MIN 13] MINISTERE DE LA DEFENSE, Livre blanc sur la défense nationale et la sécurité, Edition numérique, 2013.

[MIN 17] MINISTERE DE LA DEFENSE, Annuaire statistique de la défense, 2017.

[MON 79] MONTGOMERY C., Diversification, market structure, and firm performance: An extension of Rumelt's model, Unpublished doctoral dissertation, Purdue University, 1979.

[MON 94] MONTGOMERY C.A., "Corporate diversification", *The Journal of Economic Perspectives*, vol. 8, pp. 163–178, 1994.

[NAS 13] NASIRIYAR M., NESTA L., DIBIAGGIO L., "The moderating role of the complementary nature of technological resources in the diversification–performance relationship", *Industrial and Corporate Change*, vol. 23, no. 5, pp. 1357–1380, 2013.

[NES 05] NESTA L., SAVIOTTI P.P., "Coherence of the knowledge base and the firm's innovative performance: Evidence from the U.S. pharmaceutical industry", *Journal of Industrial Economics*, vol. 53, pp. 123–142, 2005.

[OST 10] OST, Indicateurs de sciences et de technologies, Rapport de l'Observatoire des Sciences et des Techniques, Paris, 2010.

[OUD 15] OUDOT J.-M., QUEMENER J., "Les dépenses de R&D de la BITD: une évaluation par le crédit impôt recherche", *Ecodef*, no. 74, Ministère de la Défense, 2015.

[PAL 85] PALEPU K., "Diversification strategy, profit performance and the entropy measure", *Strategic Management Journal*, vol. 6, pp. 239–255, 1985.

[PIS 00] PISCITELLO L., "Relatedness and coherence in technological and product diversification of the world's largest firms", *Structural Change and Economic Dynamics*, vol. 11, pp. 295–315, 2000.

[PIS 04] PISCITELLO L., "Corporate diversification, coherence and economic performance", *Industrial and Corporate Change*, vol. 13, pp. 757–787, 2004.

[QUI 08] QUINTANA-GARCIA C., BENAVIDES-VELASCO C.A., "Innovative competence, exploration and exploitation: the influence of technological diversification", *Research Policy*, vol. 37, pp. 492–507, 2008.

[ROB 95] ROBINS J., WIERSEMA M.F., "A resource-based approach to the multibusiness firm: Empirical analysis of portfolio interrelationships and corporate financial performance", *Strategic Management Journal*, vol. 16, pp. 277–299, 1995.

[RON 05] RONDI L., VANNONI D., "Are EU leading firms returning to core business? Evidence on refocusing and relatedness in a period of market integration", *Review of Industrial Organization*, vol. 27, pp. 125–145, 2005.

[RUM 74] RUMELT R.P., *Strategy, Structure, and Economic Performance*, Harvard University Press, 1974.

[RUM 82] RUMELT R.P., "Diversification strategy and profitability", *Strategic Management Journal*, vol. 3, pp. 359–369, 1982.

[SAV 09] SAVIOTTI P.P., "Knowledge networks: Structure and dynamics", in PYKA A., SCHARNHORST A. (eds), *Innovation Networks: Understanding Complex Systems*, Springer-Verlag, Berlin, 2009.

[SCH 82] SCHERER F.M., "Inter-industry technology flows and productivity growth", *The Review of Economics and Statistics*, vol. 64, pp. 627–634, 1982.

[SER 08] SERFATI C., "Le rôle de l'innovation de Défense dans le système national d'innovation de la France", *Innovations*, no. 28, pp. 61–83, 2008.

[SER 14] SERFATI C., *L'industrie française de défense*, La Documentation française, Paris, 2014.

[TAN 05] TANRIVERDI H., VENKATRAMAN N., "Knowledge relatedness and the performance of multibusiness firms", *Strategic Management Journal*, vol. 26, pp. 97–119, 2005.

[TEE 94] TEECE D.J, RUMELT R., DOSI G. *et al.*, "Understanding corporate coherence: Theory and evidence", *Journal of Economic Behavior and Organization*, vol. 23, pp. 1–30, 1994.

[ZYL 17] ZYLA C., Ruptures technologiques, PhD thesis, Université Paris 1 Panthéon-Sorbonne, 2017.

5

# Innovation and Legitimacy: The Case of Remotely Piloted Aircraft Systems

ABSTRACT. Building on the case of remotely piloted aircraft systems (RPAS) and the French Air Force RPAS Center of Excellence (CED), this chapter investigates how the development of innovation in defence sectors is shaped by their legitimacy, the latter being dependent on the consistency between political and technological choices.

## 5.1. Introduction

Remotely piloted aircraft systems (RPAS) are a major technological innovation for the Forces. These systems are built on a relatively simple architecture, consisting of a reusable aircraft (1) remotely controlled by a team of operators located in a ground station (2) via a radio or satellite communication link (3). They are equipped with sensors (4), sometimes effectors, and can be programmed before launch according to the mission (5). As socio-technical objects, RPAS capitalize on digital innovations related to big data exploitation, robotics and artificial intelligence. They belong to the category of complex systems and integrate a variety of payloads and sensors. In practice, their production is based on an extensive ecosystem and value chain, including a wide variety of stakeholders, organizations and communities (manufacturers, suppliers, operators, users, software developers, service companies, trainers, universities and research laboratories, insurance companies, media). The development of

---

Chapter written by Pierre BARBAROUX.

RPAS is directly dependent on the growth of services offered in fields as varied as agriculture (crop-spraying, crop management), civil engineering (mapping, construction site monitoring), transport (passengers, package delivery), health (delivery of medicines to rural or mountainous areas), surveillance and observation (borders, maritime routes, energy networks), telecommunications (transmission, relays), security (road traffic regulation, natural disaster management) and defence (combat missions, search and rescue, intelligence, surveillance, reconnaissance, and command and control, $C^2$).

While remotely piloted aircraft systems are not a recent invention[1], their development is undoubtedly linked to advances in telecommunications, space and information technology, which began in the 1950s and 1960s (SPOUTNIK was launched in 1957; ARPANET was invented in 1969; the first integrated circuit computers were used in 1963) and have continued to be used since. The growth in military and commercial applications over the past decade has been made possible due to advances in research on data collection and fusion, algorithms and artificial intelligence, data links, monitoring technologies, stealth and miniaturization. Considered a major innovation that has the potential to transform the technical skills and business models of private and public organizations in many sectors in the long term, RPAS technology used in military affairs follows a specific technological trajectory, distinct from that observed in commercial civilian markets.

---

1 Unlike civilian applications, military applications of airborne UAV systems originated at the beginning of the 20th Century. The first experiments with the use of remote-controlled or autonomous flying machines took place in France at the end of World War I, under the impetus of atypical personalities such as Captain Max Boucher. In March 1918, he carried out an experimental campaign during which he showed "that an aircraft, over a short distance, can take off, make a flight and land without the intervention of a pilot on board" ([BRU 15: 42]). The experiments continued during the interwar period, giving rise to a major invention in the history of aeronautics: the automatic pilot. The spread of this innovative technology will paradoxically put an end to the efforts made to develop remotely piloted military aircraft, but will benefit the development of commercial civil aviation. The military applications of remotely piloted or even unmanned aircraft will thus take several decades to be reborn. It is true that the prospect of seeing unmanned aircraft flying has not generated, far from it, the unconditional support of the military pilot community.

More technically mature (size, endurance, variety of sensors, altitude are variables controlled by military operators), military airborne drone systems also hold greater institutional and cognitive legitimacy than their commercial applications. On the contrary, pragmatic and moral legitimacy [SUC 95] appears weaker than in the civil and commercial sphere, due to the emotional burden that the use of armed drone systems conveys to the public. This legitimacy of innovation is key when we question the factors that favor its distribution and, through it, the development of the ecosystem that supports it. Any change, whatever its nature (e.g. technical, organizational, political), presupposes a form of acceptance that is both individual and collective, or even social, in a given cultural and regulatory environment [BAR 17]. However, the feeling of acceptance stems from the positive perception of the services offered by an innovation, for those who use it, design it or regulate its introduction into the public domain. An invention becomes an innovation not only because it is based on a new idea, but also because it creates value for any of the stakeholders in the innovation, whether it be a customer, operator, producer or user. In this perspective, the value generated by an invention depends on its legitimacy, which depends on the alignment of individual and collective perceptions expressed by stakeholders about the services it provides, with a system of rules, norms and values that characterize their environment.

In this chapter, we study the development of military applications of RPAS by considering that it is conditioned by the legitimacy they inspire. We thus suggest that the legitimacy of RPAS depends on the consistency of political decisions (regulation) with technological advances in AI, in particular those affecting the quality and safety of the services offered by these systems (autonomy, accuracy, safety of integration in airspace). In doing so, we identify the main technological and regulatory obstacles that affect the development of the ecosystem of military and civil airborne drones.

Section 2 introduces the concept of legitimacy, in all its forms, as well as strategies for legitimizing innovation. The hypothesis is that the development of innovation is a process of legitimization based on the implementation of deliberate strategies and policies. Section 3 examines the case of RPAS in France through two complementary dimensions: regulatory and technological. These dimensions are the main obstacles to the development of the legitimacy of military (as well as civilian) drones. In this context, the role played by the French Air Force RPAS Center of Excellence

(CED) in terms of legitimization strategy will be analyzed. Section 4 recalls the main results of the study and opens up questions for future research.

## 5.2. Technological innovation and legitimacy

The ability to influence perceptions and representations is essential to the legitimacy of an innovation. In practice, legitimacy reflects the acceptance, understanding, adherence to, and even support of the various stakeholders involved in an innovation. In this context, the innovation must not only *convince* them of its legitimacy in relation to their interests and needs, but also appear to *comply with* the norms, rules and values that characterize the host environment [ROC 01]. Two aspects of legitimacy are thus highlighted: institutional and strategic [SON 10]. The institutional vision considers legitimacy as a result of the socialization process. From this perspective, the innovation whose legitimacy is assessed is somehow "selected" by the environment because of its conformity with the system of rules, norms, values and beliefs attached to that environment. Conversely, the strategic vision considers the legitimacy of an innovation as the result of a deliberate plan to achieve consensus on the conformity of its action and the services it offers with the norms and rules characteristic of its environment [SON 10]. Here, legitimacy is instrumentalized and perceived as a resource by actors capable of "*manipulating and deploying symbols*" likely to create the conditions for a convergence of perceptions [SUC 95: 572].

In this context, the challenge is to understand how an innovation gains legitimacy. [BAR 17] has shown that the process of legitimizing an innovation (organizational in this case) involves the implementation of a set of actions that facilitate the acquisition, maintenance and development of legitimacy. [ZIM 02: 415] identify four categories of action or legitimization strategies: conformation, selection, manipulation and creation. Conformation involves, for innovation, respecting the rules imposed by the system by acting "*in accordance with the demands and expectations of the existing social structure*" [ZIM 02: 422]. Selection, for an innovation, consists of choosing the socio-economic environment in which it should operate. For example, it may be chosen to develop some applications of airborne UAV system technology (e.g. military intelligence), rather than others (e.g. passenger transport). Manipulation implies "*changing the environment*" in order to ensure greater coherence between the services provided by innovation and its use environment [ZIM 02: 424]. This strategy very often involves the implementation of communication policies by innovation stakeholders. Finally, the purpose of creation is to create a normative and regulatory environment conducive to increasing the legitimacy of innovation. This is the case with breakthrough innovations such as automobiles, whose introduction has justified heavy and costly investments (e.g. signaling, road network development), partly initiated (and financed) by private stakeholders (e.g. Michelin invented road signs at the beginning of the 20th Century in France) and aimed at creating the infrastructure necessary for the adoption of an innovative individual and public transport mode.

The previous definition of legitimacy as a process suggests that it results from intentional strategies on the part of actors. These strategies condition the emergence and maintenance of different types of legitimacy, each linked to a different source of legitimacy. [SUC 95] distinguishes three sources of legitimacy: pragmatic, moral and cognitive. Pragmatic legitimacy results from a "*rational calculation*" by individuals belonging to the immediate innovation environment [SUC 95: 578]. It can be broken down into three types. It may thus be rational for a group of individuals to consider an innovation to be legitimate whose services (i) are favorable to them, (ii) are in the interests of the group or (iii) are motivated by values shared by the group. Moral legitimacy more accurately reflects "*a positive normative evaluation of the entity* [innovation, *added by us*] *and its activities*" [SUC 95: 579]. [ALD 94] and [ZIM 02] describe this institutional, normative and regulatory source of legitimacy as socio-political. It can be of four types and result from a social judgment about (i) the services provided by innovation, (ii) the procedures and techniques it mobilizes to deliver services, (iii) its ability – understood as an organized system of procedures – to provide certain services or (iv) the individual qualities of the individuals who embody it [SUC 95: 579-581]). Finally, cognitive legitimacy [SUC 95: 582] can be of two types and can be related to the ability of actors to understand innovation and the services it provides (understandability) or to consider them objectively legitimate (*de facto* legitimacy). [ALD 94], in their study of the influence of legitimacy on the emergence of new industries, consider that cognitive legitimacy is based on the shared knowledge between stakeholders of the new industry, the products it offers, their functions, their value and their uses. This source is therefore important in understanding the innovation and the services it offers. Table 5.1 summarizes and presents the different sources of legitimacy of an innovation, as well as the types and strategies associated with it.

| Sources of legitimacy | Types of legitimacy | Elements of definition | Dominant strategies |
|---|---|---|---|
| Pragmatic | Exchange | Supported by the rational interest of agents in innovation and related services | Conformation<br>Selection<br>Manipulation<br>Creation |
| | Influence | | |
| | Provisions | | |
| Moral/socio-political | Consequences | Based on the normative evaluation of the quality of services provided by an innovation | |
| | Procedural | | |
| | Structural | | |
| | Personal | | |
| Cognitive | Comprehensibility | Resulting from the *de facto* understanding or acceptance of innovation and the value it creates through the services it delivers | |
| | *De facto* legitimacy | | |

**Table 5.1.** *The forms of legitimacy of innovation (according to [ALD 94, SUC 95, ZIM 02] – adapted from [BAR 17])*

## 5.3. The ecosystem of RPAS in France

France is a leading nation in the production, operation and regulation of civil airborne drone systems. In 2012, the French Civil Aviation Authority (DGAC) was one of the first government agencies to define a regulatory framework aimed at codifying the professional uses of airborne drone systems. In this context, RPAS producers and operators have a legal obligation to declare to the DGAC the type of aircraft produced and/or used, the nature of the activity and/or mission, as well as the scenario in which the activity and/or mission is used, in order to be issued a certificate of airworthiness. In addition, operators and users of captive and non-captive RPAS must have been trained, be able to provide a certificate of completion of training issued by the DGAC or hold a certificate of fitness for remote pilot duties (note that theoretical training for remote pilots came into force in July 2018).

Table 5.2 presents the seven categories of airborne drone systems listed by the DGAC, as well as the four scenarios of use identified.

| **Categories** | | | | | | |
|---|---|---|---|---|---|---|
| **A** | **B** | **C** | **D** | **E** | **F** | **G** |
| Air-motorized or non-motorized models <25 kg<br><br>Inert gas aircraft <25 kg | Aeromodels different from A | Captive unmanned aircraft <150 kg | Unmanned, non-captive aircraft <2 kg<br><br>Inert gas aircraft <2 kg | Unmanned aircraft, different from C or D <25 kg<br><br>Inert gas unmanned aircraft <25 kg | Unmanned aircraft, different from C, D or E <150 kg | Unmanned aircraft, different from C, D, E, F |

| **Scenarios** | | | |
|---|---|---|---|
| **S1** | **S2** | **S3** | **S4** |
| Demand transactions<br><br>Unpopulated and populated areas<br><br>Max. weight: 25 kg<br>(Altitude: 150 m)<br>(Distance: 100 m) | Out-of-sight operations (<1 km)<br><br>Unpopulated areas<br><br>Max. weight: 25 kg<br>(Altitude: 50 m)<br>(Distance: 1000 m) | Demand transactions<br><br>Unpopulated and populated areas<br><br>Max. weight: 8 kg<br>(Altitude: 150 m)<br>(Distance: 100 m) | Out-of-sight operations (>1 km)<br><br>Unpopulated areas<br><br>Max. weight: 2 kg<br>(Altitude: 50 m)<br>(Distance: 1000 m) |

**Table 5.2.** *Categories of unmanned aircraft and typology of employment scenarios (source: DGAC, Fédération professionnelle de drone)*

Over the past ten years, professional applications of airborne drone systems have been booming. In 2013, the DGAC listed 40 companies producing airborne drone systems in France. This figure has more than quadrupled in four years to reach 170 companies at the end of 2017, representing an average annual growth rate of 53%. The French company Parrot is one of the two world leaders in the production of all types of airborne drone systems, along with the Chinese company DJI. For RPAS operators, the growth rate is even more spectacular. In 2013, the DGAC registered nearly 400 civil operators (for a total of 1400 in Europe). In 2017, there were more than 6700 of them, representing an average annual growth rate of nearly 131%. Figure 5.1 shows the evolution of the number of producers and operators of airborne drone systems in France from 2013 to 2017.

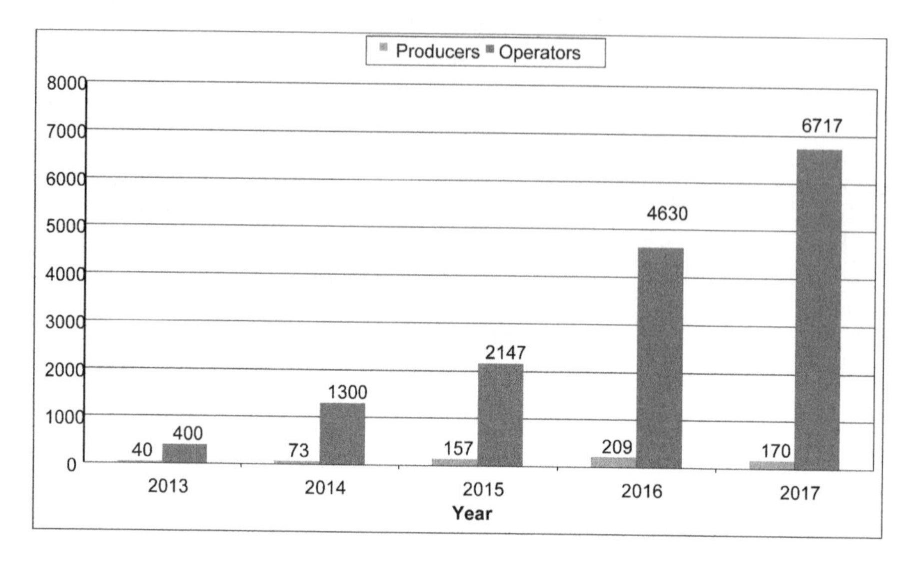


**Figure 5.1.** *Evolution of the number of producers and operators of RPAS in France (source: DGAC, Fédération professionnelle du drone, Ministère de l'écologie et des transports). For a color version of the figures in this chapter see www.iste.co.uk/barbaroux/technology.zip*

Despite strong growth in services and commercial activities, the majority of civil operators declared and holding a certificate of airworthiness are very small enterprises (VSEs) using a small number of category C, D and E aircraft (90% of unmanned aircraft have a total loaded weight of less than 10 kg) in scenarios corresponding to "visual" operations (S1 & S3). In addition, more than 98% of operators registered with the DGAC offer observation-related services [OBS], including the training of remote pilots for this type of mission [FOR-OBS]. The sectors of activity concerned are mainly audiovisual and media (more than 50%), building and infrastructure inspection (20%) and civil engineering (15%). Few operators in any industry are able to offer multiservice solutions involving the use of endurant aircraft, operated beyond the visual line of sight over long distances, and capable of carrying several types of sensors. However, sectors such as agriculture, security and surveillance express needs in this direction (agriculture is the sector with the greatest medium-term growth potential for airborne drone systems).

While military RPAS have an architecture comparable to that of civilian systems (e.g. aircraft, ground station, data link, sensors), their definition also includes a deployment and recovery system, training and education capabilities and a support system for maintenance, airworthiness and human resource (HR) management. As in the civilian sector, the classification of military airborne drone systems is based on the criteria of mass, range ("visible" and "out of sight"), endurance and altitude of the aircraft, and the type of mission for which it is used. There are three levels of employment (strategic, operational, tactical) as well as small, mini and micro drones. In France, the classification of military RPAS is based on three categories. Contact drones (category I, tactical use) have a mass of less than 150 kg and are remotely controlled visible (e.g. DRAC). Tactical drones (category II, operative use) have a mass between 150 and 600 kg and are also visually controlled (e.g. Watchkeeper). Strategic and attack drones (category III) have a mass of more than 600 kg and are unmanned and controlled via satellite links (e.g. Reaper, Predator, Harfang).

The development of applications for airborne drone systems, whether civil or military, depends on several factors. Experimentation and certification campaigns by aviation regulatory authorities (e.g. the DGAC in France, the Federal Aviation Administration, FAA, in the USA) can accelerate or hinder this development, so do governmental industrial policies. Public opinion, media coverage and privacy issues are also likely to influence the growth of the airborne drone ecosystem. Finally, advances in airspace management and air safety, including embedded systems such as sense-and-avoid algorithms, can facilitate the integration of pre-programmed or even autonomous RPA into general air traffic (GAT) and encourage their dissemination. Together, technological, regulatory and public opinion factors contribute to the legitimacy of airborne drone systems in many forms. While military applications enjoy *a priori* greater socio-political and cognitive legitimacy than civilian applications, they are nevertheless dependent on the state of public opinion regarding the use of drones for purposes other than surveillance, intelligence, communication or command and control of operations ($C^2$). In addition, aircraft airworthiness constraints apply to all types of aircraft, whether state or non-state, commercial or non-commercial, manned or unmanned flight safety and air safety issues are shared between civilians and the military and play a decisive role in the development of airborne drone system applications.

Because they share a common regulatory framework (e.g. airworthiness rules) and common concerns for flight safety and security, that they depend on technological advances in dual knowledge areas and that they are highly sensitive to public opinion (and to any event likely to influence it), it seems difficult to establish a clear division between the sources of legitimacy of military and civilian airborne drone systems (we are aware, however, that some armies in the world are reluctant to use of “killer drones” because of public opinion). In order to understand what strategies are being implemented by military organizations to promote the legitimacy of innovation from airborne drone systems, the following section examines the case of the Drone Center of Excellence deployed by the French Air Force at École de l’Air in Salon-de-Provence (Air Base 701).

## 5.4. The role of the French Air Force RPAS Center of Excellence (CED) in legitimizing RPAS systems

Created and located since September 2014 at Air Base 701 (BA 701) in Salon-de-Provence, the RPAS Center of Excellence (CED) is the aggregation of the French Air Force’s expertise in the field of airborne drone systems. With the support of the Office Nationale d’Etudes et de Recherches Aérospatiales (ONERA, French National Aeronautics and Space Agency), the French Air Force Research Center (CReA) and the International Academic and Aeronautical Training Center (CIF-AA) of the company Défense Conseil International (DCI), the CED offers its partners a wide range of services oriented towards research and experimentation, the design and evaluation of airborne drone systems and the training of drone remote pilots.

The CED is composed of about 10 personnel, officers and non-commissioned officers, in charge of leading the various activities, including (1) training drone pilots within a joint and inter-ministerial framework, and (2) technological experimentation through, in particular, the production of proof of concepts related to the expression of specific operational needs (state or non-state, military and civilian).

From the outset, the CED has been recognized for its dual scientific and operational culture. Its director, a doctor in computer science and an officer in the French Air Force, developed, during his doctoral research at the French Air Force Research Center, intelligent algorithms enabling a remotely piloted airborne system to "sense and avoid" any obstacle that might arise in the airspace. His research on artificial intelligence (AI) has been the subject of several patent applications. Its presentation at conferences and research seminars attracted great interest both from researchers and aerospace and defence (A&D) manufacturers. This research has mainly enabled the CED to gain visibility and credibility within the community of researchers and drone operators (e.g. the CED is a member of the *RPAS network* of the SAFE competitiveness cluster of the PACA region). They have also positioned the Air Force as a stakeholder capable of actively participating in the training of state aircraft remote pilots (military and civil) and, more broadly, in the development of the airborne drone ecosystem.

When the CED was inaugurated in 2014, there was a huge need for certification of drone operator training, definition of the regulatory frameworks governing their use, operational testing and experimentation, and research and development. Rather than undergo the technological and regulatory transformations, the Air Force has therefore chosen to act by investing in the creation of a new unit headed by an officer with an atypical profile: an expert in military aeronautics and a professor-researcher in the fields of artificial intelligence (AI) and automation.

The roadmap for the new unit has thus been established around two axes: the training of drone pilots and experimentation in the service of innovation. Rapidly, its director proposes to translate this roadmap by structuring the center's activities around the following four missions:

– produce training syllabuses for mini-drone pilots (mass less than 25 kg) and provide training for state and private trainees. The final objective, beyond satisfying the need for training of French Army RPAS pilots, is to establish the CED as a partner of the DGAC in the certification of theoretical training for remote pilots of categories D and E aircraft;

– deliver high-level training aimed at raising awareness among the economic fabric (SMEs and SMIs) and key account managers of the potential of airborne drone systems, either by integrating their technologies into remotely piloted or autonomous airborne systems or by using drone systems in their respective sectors of activity;

– design and implement, in partnership with DCI, drone pilot learning and training systems based on the simulation of environments associated with *all* types of civil and military RPAS;

– develop the Air Force's experimental capabilities in the context of strengthening air safety and aircraft autonomy, through the design of various remote-controlled and autonomous platforms and their associated embedded algorithms, in particular in the fields of object recognition, navigation and classification.

Each of these missions undertakes actions that, in turn, generate or reinforce the development of different forms of legitimacy attached to airborne drone systems. The training of drone pilots of light airborne drone systems (all types of scenarios, S1–S4) contributes to establishing the legitimacy of technology in its pragmatic and cognitive forms by providing expertise in technical and legal fields, which in turn contributes to the spread of a "drone" culture within diversified professional circles (e.g. military, gendarmerie, civil security, state operators). In the same vein, economic actors who attend high-level training courses deepen their understanding of the services provided by airborne drone systems, perceive their value through the multiple potential applications presented and discussed during the training, and even allow airborne drone systems to achieve *de facto* legitimacy.

The assistance provided to the DGAC in the field of training certification supports the development of moral and socio-political forms of legitimacy by contributing to the creation of an institutional environment conducive to the development of airborne drone system applications. The interest for the French Air Force and, through it, for all its partners, to participate in the process of producing standards for the training of remote pilots in France, is to minimize the risk of non-compliance of military drone pilot training with the civil aviation normative system which, as for the airworthiness of state aircraft, applies to all [SGD 16]. The general air traffic rules (Article D 131-3 of the Civil Aviation Code) are binding on pilots of aircraft operating in military air traffic (defined by order of the Minister of the Armed Forces; Article D 131-6) and on all military air traffic service providers (Article D 131-8).

Finally, activities to design learning systems through simulation and to develop platforms and algorithms contribute to improving the legitimacy of airborne drone systems in their moral and socio-political dimensions. Simulation and experimental research are an essential contribution to improving flight safety and the integration of airborne drone systems into the airspace. These activities make it possible, on the one hand, to develop quality operator training and, on the other hand, to encourage experimentation on innovative concepts in a healthy and safe environment. This last point is particularly sensitive insofar as the Air Force is solely responsible for the permanent surveillance of national airspace (i.e. the permanent air safety posture, PPS-Air). However, the detection, identification and possible neutralization (interception) of airborne drone systems with a total mass of less than 25 kg is difficult or even impossible, as civilian and military radars distributed in a network throughout the national territory are ineffective due to the small size and reduced speed of these aircraft. In the absence of a clear radar signature, only visual detection is effective but limited, especially at night. The CED is thus working to reduce this capacity gap and is experimenting with various innovative concepts in the detection, identification and interception of micro and mini drones. These experiments are complementary to the research conducted at Salon-de-Provence (respectively by CReA and ONERA) on active and passive radars. Some of these exploratory experimental works lead to the production of proofs of concept that can be the subject of subsequent technical-operational developments. For others, the experiences conducted at the CED demonstrate their low development potential, given the current state of knowledge and technology (i.e. “killing bad ideas”).

## 5.5. Implications and conclusion

The example of the CED suggests that the development of airborne drone systems, considered as a disruptive technological innovation in many sectors, depends on the implementation of actions aimed at developing their legitimacy, in all its forms. In particular, we show in this chapter that the CED has chosen to combine three legitimization strategies within the scope of [ZIM 02]: conformation, selection and creation. In practice, the implementation of the first strategy has conditioned the implementation of the selection and creation strategies. The desire to comply with general air traffic regulations justified the choice to position the Center in the field of training and experimentation for operational purposes, applied to a particular category of airborne drone systems: those with a total mass not exceeding 25 kg. This category represents the majority of unmanned aircraft currently registered by the DGAC (commercial professional use) or employed in the Forces for military purposes (e.g. intelligence, surveillance, reconnaissance). This choice, consisting of selecting a specific aircraft category covering the four employment scenarios (S1–S4), then justified the CED's focus in terms of training. The center first produced the syllabuses and delivered the knowledge needed to train civil and military remote pilots. Then, by obtaining certification of its training syllabuses by the DGAC, the CED became involved in structuring the regulatory environment governing the use of airborne drone systems, making it conducive to the development of civil and military applications for small unmanned aircraft. At the same time, the CED opted for experimentation applied to the development of technological blocks essential for flight safety and air safety (e.g. AI and system autonomy) and facilitated the integration of small airborne drones into general air traffic. Together, the selection and creation strategies implemented by the CED have helped to ensure that the services offered by airborne drone systems are consistent with the existing, albeit incomplete, system of standards and rules.

It is therefore through the combination of legitimization strategies and their impact on perceptions and the regulatory environment that an invention such as airborne drone systems becomes an innovation. On this point, the case study shows that the strategies implemented by the French Air Force through the CED are not only complementary, but also developed over time. This observation is in line with the objective of researchers and practitioners to precisely understand "*how strategies can be combined to build legitimacy*" [ZIM 02: 428]. Thus, in this case, the selection precedes the creation of an environment conducive to the development of airborne drone system applications, the latter being necessary due to the lack of a mature regulatory framework to comply with. We then see different stages in the process of developing the legitimacy of an innovation, each supported by an adapted strategy. These steps, arranged in different temporal sequences, define a process of legitimacy evolution. While we do not observe a clear correspondence between a strategy, a phase and a source of legitimacy, the example of the CED shows, however, that pragmatic and moral sources of legitimacy are essential in the initial stages of the legitimization process. In practice, the cognitive legitimacy of innovation (i.e. the understanding and *de facto* acceptance of airborne drone systems) depends on the positive assessment by the various stakeholders of the services provided by the new technology. This evaluation is carried out with reference, on the one hand, to their personal interests (pragmatic source) and, on the other hand, to shared collective values (moral source of legitimacy). It can therefore be observed that pragmatic and moral sources of legitimacy are associated with the initial phases of the process, namely "*building and perpetuation of legitimacy*" [BAR 17: 57]. Once the legitimacy support base has been established, it can then be developed. For example, the DEC's investment in the deployment of simulation-based learning tools, initially limited to the training of small drone pilots, has broadened the scope of training to include larger drone pilots, as they are more aligned with military capabilities and operational requirements.

The analysis developed in this chapter obviously has limitations since it is a case study of a particular type of technological innovation operating in a particular professional environment. The representation of the process of legitimizing the proposed innovation therefore does not constitute a model from which to generalize the purpose. However, it does draw the attention of researchers and managers to the role played by legitimacy, in all its forms, and the legitimization strategies that can be implemented to transform an invention into an innovation.

## 5.6. References

[ALD 94] Aldrich, H.E., Fiol, M.C. (1994), "Fools rush in? The institutional context of industry creation", *The Academy of Management Review*, vol. 19, no. 4, pp. 645–670.

[BAR 17] Barbaroux, P., Gautier, A. (2017), "En quête de légitimité: la gestion du changement organisationnel comme processus de légitimation", *Management International*, vol. 21, no. 4, pp. 48–60.

[BRU 15] Brun, C. (2015), "Max et les "ferrailleurs" ou l'histoire inachevée de l'avion sans pilote", in Mazoyer, S., Lespinois, J., Goffi, E., Boutherin, G., Pajon, C. (eds), *Les Drones Aériens: Présent, Passé et Avenir. Approche Globale*, La Documentation française, pp. 39–63.

[ROC 01] Rocha, R., Garnerud, L. (2001), "The search for legitimacy and organizational change: The agency of subordinated actors", *Scandinavian Journal of Management*, vol. 27, no. 3, pp. 261–272.

[SGD 16] Secrétariat Général de la Défense et de la Sécurité Nationale (2016), L'essor des drones aériens civils en France: enjeux et réponses possibles de l'Etat, Rapport du Gouvernement au Parlement, p. 62.

[SON 10] Sonpar, K., Pazzaglia, F., Kornjenko, J. (2010), "The paradox and constraints of legitimacy", *Journal of Business Ethics*, vol. 95, no. 1, pp. 1–21.

[SUC 95] Suchman, M.C. (1995), "Managing legitimacy: Strategic and institutional approaches", *The Academy of Management Review*, vol. 20, no. 3, pp. 571–610.

[ZIM 02] Zimmerman, M.A., Zeitz, G.J. (2002), "Achieving new venture growth by building legitimacy", *The Academy of Management Review*, vol. 27, no. 3, pp. 414–431.

PART 2

# Transformation of Skills and Uses Induced by Innovations

# 6

# Man–machine Teaming: Towards a New Paradigm of Man–machine Collaboration?

ABSTRACT. The integration of automation in aircraft cockpits had raised questions about human–system collaboration in its time. The concept of shared authority has led the human operator, abusively considered as the weak link in aviation safety, to move away from the decision-making loop in favor of support systems. The emergence of artificial intelligence-based help systems has brought this human–machine collaboration theme to the forefront. This time, the so-called "intelligent" systems spread their superiority over the human being in many fields and question de facto the relevance of continuing to imagine any kind of collaboration.

## 6.1. The challenges of collaboration

> "Getting together is the beginning, keeping together is progress, working together is success." Henry Ford

The proliferation of AI and big data applications raises the question of the use of these technologies by Armed Forces in future commitments. These technological opportunities are of particular interest to Command and Control ($C^2$) structures and raise fundamental questions about command organization and practice. The computational power of the latest generation of computers combined with the volume of data accessible through cyberspace provides an unprecedented flow of information. The processing of the growing amounts of data accessible by the Armed Forces is a new operational opportunity. The operational implementation of AI and big data technologies questions the presence of humans in the loop. Are we engaged

---

Chapter written by Vincent FERRARI.

in a new collaboration between human agents and artificial agents or is the purpose of AI-based decision support systems to take us out of the decision-making loop (see Box 6.1 for the opinion on this issue from the former Chief of Staff of the French Air Force, General Denis Mercier)?

In aviation, the integration of automation in cockpits was the first step towards collaboration between human agents (pilots) and support systems (automation). The concept of shared authority is the theoretical basis for this collaboration. After more than 50 years of integrating automation in aviation, what have we learned from human–machine collaboration? Can the feedback from the sharing of authority enable us to anticipate the challenges that will have to be resolved when implementing collaboration between a human agent and an "intelligent" artificial agent? But is artificial intelligence only intended to collaborate with humans? Isn't the objective of AI, especially strong AI, to replace human in certain tasks, or even in all of them?

"I don't think we can say that we should always keep humans in the loop. Having humans in the loop is needed for decisions regarding the use of force or decisions regarding command of operations. But autonomous systems are also logistics systems that do not necessarily require human control. An automatic intervention by an autonomous robot following the detection of a serious injury will be even more effective if it frees itself from lengthy decision-making processes that people would make in the loop. Too often, we insist on the need to keep people in the loop because we associate autonomous systems with killer robots, but this is far too restrictive.

Cyber is another area where we will not have time to place people in the loop to disconnect a network that is under significant attack, for example. Moreover, people in the loop do not necessarily mean that there must be permanent control of the use of force. A robotic system that could enter a cave where terrorists and families or hostages are mixed and decide on its own on the use of force based on advanced autonomous detection systems is possible. But it will be up to humans to decide their tactical use according to the situation and, in a way, it is another way to involve humans in the loop even if significant autonomy is deliberately left to the machine in certain situations. Finally, the question arises, with autonomous systems based on digital architectures, regarding the degree of human involvement in the loop. These architectures are relevant if they allow more responsive operational organizations that promote the devolution of operational control of forces".

**Box 6.1.** *Opinion of General Denis Mercier (former Chief of Staff of the French Air Force) on the presence of human in autonomous systems (personal conversation)*

## 6.2. The sharing of human–machine authority: the premises of collaboration

In the field of civil and military aeronautics, the problems inherent in the implementation of effective collaboration between a human agent and a support system have been addressed in the context of automation and authority sharing.

Computer scientists have been automating transport aircraft since the 1930s. This automation has used analogue and then digital techniques to switch to symbolic techniques and tools. Aircraft cockpits have evolved from electronic equipment to computer equipment. In the 1980s, the flight management system (FMS) was introduced as a database management system on board commercial transport aircraft. It freed the crew from flight planning and navigation tasks because this system included the functionality necessary to produce results that pilots were used to generating themselves. The underlying methods were rule-based systems and trajectory optimizations.

Authority sharing, which is the theoretical concept supporting automation, embodies a strategy for allocating the resources of two types of agents – human and the support system – according to a given task, a specific activity. In civil aeronautics, given the problems of air traffic flow management and responsibilities related to the transport of civil passengers, there is little room for a human decision based on approximations (i.e. heuristics, decision-making biases). The vast majority of authority belongs to pilot support systems; pilots manage communications and supervise the actions (decisions?) undertaken by systems. In military aeronautics, given the "dynamic" nature of the environment, adaptation, or even improvisation, inherent in human decision-making is still key for the time being. Between civil and military aeronautics, the concept of authority sharing is therefore the same, but its actual implementation (so far) is different.

Thus, the objective of authority sharing is to identify, according to their "competencies", the respective roles of humans and systems. Once the evaluation is complete, it is necessary to define the level of automation (for the system), the nature of the training (for the human) and design the appropriate human–system interfaces. There are several levels of automation, from "fully automated" to "fully manual" and between these on this continuum, to bring out effective collaboration; it is still necessary to determine what the two agents need to know about each other's action. In theory, the evaluation of human knowledge about the functioning and capabilities of the support system should allow the development of “fluid”, “intuitive” interfaces. These “innovative” interfaces were to provide the human agent with the ability to control and understand the actions carried out by automatisms (i.e. supervision) while maintaining a sufficiently high level of vigilance to enable people to intervene effectively in the event of a system failure (i.e. make a decision). Without of course questioning all the benefits, in terms of flight safety, that automation brings, it is not exaggerated to put forward the idea that the sharing of human–machine authority does not keep all its promises.

First, the “sharing” of authority is largely in favor of automatisms and at the expense of the skills of human agents. Indeed, the current trend in the field of authority sharing is to minimize, as much as possible, the intervention of the human agent considered, in a detrimental way, as the “main source of errors”, an “impassable air safety threshold”. Moreover, understanding human skills is too complicated and human behavior models are not detailed enough for computer scientists, who are more comfortable with algorithms than with human heuristics.

Second, no specific training approach has been initiated to explain to the human agent how support systems are designed, what their advantages and limitations are, how to avoid the pitfalls associated with automatisms and on what theoretical principles aid systems have been developed. The human agent therefore has no idea how a system that is supposed to help him works. It should be noted here that when a “collaboration” between a human agent and a support system is made difficult because the system is complex and the human agent is considered unreliable, then it is often simpler to ask the “weak link” to undergo training that requires an additional effort through demanding, often time-consuming and costly training.

Third, the interfaces developed under the principles of authority sharing are neither “fluid” nor “intuitive” but still reflect the complexity inherent in help systems (e.g. many windows to open, drop-down menus). Precisely, the ergonomics of the interfaces have not yet been able to make the user aware of what the system is doing, what it can do and especially what it cannot do. Unfortunately, even today, the ergonomic approach is still too often implemented during user testing phases, when systems are designed and already almost operational. The approach then reduces its scope to an improvement in the figurative representation of information on displays and other interfaces. This "opacity of systems" makes it impossible to implement efficient human–system collaboration and becomes catastrophic when systems are inoperative (i.e. what is commonly referred to as the "cliff effect" observed, in particular, during the AF447 disaster).

Finally, the sharing of authority implies a role of supervisor and decision-maker for the human agent, whereas we observe a disinvestment of the latter under the pretext that automatisms effectively manage routine situations. In other words, when the sharing of authority is too favorable to support systems, human agents feel less and less concerned by system actions/decisions (e.g. this loss of motivation has consequences on the quality of the attention paid by pilots to their activity) and lose the skills that the systems implement in their place. This phenomenon is so frequent that it has been called "the irony of automation" by security specialists. Between the opacity of systems and the irony of automation, the human agent will quickly tend to delegate what remains of his authority to the support system in a relationship of dependence, and even submission. This dependence on systems is not problematic in the vast majority of cases since they effectively manage routine situations. In the case of rare situations, which fall outside the "competence" domain of automated systems, the latter will (re)support a human agent who is faced with a critical situation, completely disconnected from the situation, with no support system at his disposal and too little time to understand and act! Under these conditions, if the human agent fails to find a way out of the crisis, then he will automatically be designated as the weakest link in air safety.

In summary, the concept of authority sharing implies a division of tasks according to the respective competences of human and system to achieve an optimal level of collaboration. The trade-off in favor of support systems can be explained either by the operator's unreliability (from an industrial point of view) or by the inability of automation engineers to accurately identify the human operator's capacities and to make the field of expertise of each of the collaboration's protagonists "readable" (from the perspective of human factor specialists). While in a large majority of cases, air accidents are therefore the consequences of a failure in collaboration between human and support systems, these situations remain, fortunately, infrequent, thanks in particular to more than half a century of experience and feedback in the field of automation integration. But today, with the rise of artificial intelligence, support systems are becoming increasingly autonomous and the notion of human–machine collaboration (and consequently that of authority sharing) is being called into question. How to ensure efficient collaboration between human and artificial agents? Have AI specialists drawn on the experience gained through automation and authority sharing to develop their research effort towards understanding human behavior (i.e. integration of cognitive models into AI) and the readability of artificial agent functioning (i.e. development of more intuitive human–system interfaces)?

## 6.3. Expert systems and human–system collaboration

The concept of authority sharing, developed in both civil and military aviation, has not been able to exploit the full potential of effective collaborative work between a human agent and a support system. Sharing authority is more like juxtaposing two sets of skills, two different methods for understanding the complexity of the world (e.g. calculation for the support system and adaptation for humans). The human agent has had to and has been able to adapt to this "centered engineering" approach in terms of collaboration. Will the new integration of artificial intelligence-based support systems into aircraft or command and control systems of the future change the nature of human–system interactions? Before attempting to answer this question and determine whether an AI agent is "capable" of collaborating with a human agent, it seems important to briefly recall what a collaborative activity involves.

Mucchielli identifies the conditions necessary for the success of collective work for a group of human operators [MUC 17]. Among these, the following conditions are transferable in the context of collaboration between human and artificial agents:

– unambiguous communication between agents;

– mutual assistance between all agents;

– the willingness of a default member to replace him/her;

– knowledge of the skills and limitations of each of the agents involved in the collaboration (this point is particularly developed in this chapter);

– the division of labor after the definition of common objectives;

– the combination of forces.

To understand human skills you need to know how they work. Intelligence is a characteristic of higher living organisms, particularly humans, the study of whom being the responsibility of psychology. Until the 1960s, many theories developed by cognitive psychologists could be used by computer scientists, future specialists in AI (e.g. Gestalt's work on human vision). The first effective collaboration between support systems and people dates back to the 1980s with the advent of expert systems. Expert systems are computer programs that integrate a knowledge base (often extracted from that of a human expert) used as a database by a system of logical inferences. With expert systems, the problems have focused on the question of explicitly representing human knowledge. The theoretical model that structures expert systems is therefore a model derived from human behavior (that of an expert) classically studied by cognitive psychology [FER 16, 18, GOB 00, SIM 96, 97]. The development of expert systems embodies not only a very strong collaboration between the cognitive sciences and informatics but also the beginning of a "paradigmatic gap" that will widen between these two fields. On the one hand, computer scientists have found an effective paradigm (knowledge-based systems) which, thanks to big data, can be applied to a large number of fields. On the other hand, cognitive researchers no longer provide sufficiently detailed behavioral models to guide computer scientists' research. Indeed, under the impetus of technological advances in brain imaging, cognitive sciences are more interested in the functioning of the brain than in human behavior. Unfortunately, brain activity visualization technologies are not yet mature enough to allow researchers to propose robust and valid models. The external validity (i.e. the generalizable nature) of a phenomenon observed in a scanner is, by definition, very low. For these reasons, everyone returned to their field of specialization with very little regret for interdisciplinarity difficulties.

## 6.4. AI and collaboration between human and artificial agents

Models of human functioning provided by cognitive psychology (including expertise models) combined with the increasing computational power of computers have contributed to the emergence of artificial intelligence (AI). The challenge of AI is to successfully equip machines with the capabilities to perform tasks or activities that are considered "intelligent" (in the sense that they have so far only been performed by humans). Classically, a distinction is made by specialists between low and high AI.

### 6.4.1. *The omnipresence of weak AI*

Artificial intelligence is said to be weak when it only reproduces a specific behavior but not its functioning. Among the programs developed on the principle of weak AI, there are mainly expert systems (knowledge-based systems), software capable of simulating the behavior of a human performing a specific task, for which he is the only one with the know-how (e.g. reading radiographs, playing chess, etc.). The fields of application of weak AI are numerous, the most mediatized being autonomous cars, the game "go", face recognition, analysis of medical radiographs, allocation of bank loans, or more recently the detection of lies during affidavits. GAFA (Google, Apple, Facebook and Amazon) also implement many weak AI systems to improve our user experience, track our consumption habits or make us as dependent as possible on the latest application available on any social network. It is weak AI that sorts our emails, appointments, personal and professional information on our smartphones, photos, with our agreement and our more or less conscious participation.

When used in very complex domains that require enormous computational power to process large amounts of data, weak AI systems are extremely effective and, in most cases, useful to our society. Since Turing [TUR 50], machine intelligence has been primarily a matter of computational power and the ability to learn from data. For the author, the problems related to the representation of human knowledge have no place in the field of artificial intelligence. The operative basis of weak AI is based on the recognition of forms, situations, which would correspond to the implementation of so-called "low-level" processes in humans, such as perceptual processes. These processes are traditionally distinguished from so-called "high-level" processes such as reasoning or decision-making[1].

### 6.4.2. *The opacity of weak AI*

Today, weak AI is omnipresent; it works for us, sometimes with us, often in our place. Depending on the application scope of weak AI systems, humans are either considered as supervisors or consumers. Weak AI programmers do not seek to emulate the functioning of human beings; they

1 It should be noted that this distinction is now a little obsolete given the interest that cognitive psychology research has in perceptual processes.

have not thought and developed their technology with human–system collaboration in mind or even in that of making the functioning of their system accessible to the understanding of users. Since the cognitive sciences no longer provide human behavioral models, AI specialists have been programming "intuitive" models of intelligence. In the development of weak artificial intelligence, it is not necessary to understand how human beings think. Aircraft fly and yet they do not flap their wings. Aeronautical engineers have preferred another approach to flight, less "behavioral" and technically more accessible, but which in the end has little to do with the flight of a bird. Programmers have therefore defined systems capable of reproducing certain human capacities (mainly perceptual) without reproducing the cognitive processes intrinsic to these capacities. However, this independence from cognitive sciences is such that there has been a change within AI itself [VAN 08] to remind programmers that they cannot always be satisfied with their intuition to judge the "good" behavior of a formal model, and that AI formal theories must be constrained by the empirical facts established by the psychology of reasoning and decision [PEL 05, PIE 03]. Moreover, the improvised models invented by computer scientists are combined with the convoluted complexity of increasingly sophisticated algorithms (deep learning is an example of this enmeshment of calculations) resulting in systems whose functioning has become almost inaccessible to the programmers themselves[2]. Finally, human–system collaboration is more complicated than it was when using automatic systems and we are far from the idea of "mutual knowledge of the skills and limits of each of the agents involved in collaboration".

Box 6.2 presents a summary of a recent upstream study project (PEA) entitled "the man–machine teaming". Despite a promising title when we look at human–machine collaboration, we can be surprised that the collaborative work described in this project focuses on the monitoring of operators, i.e. on the monitoring of the proper "functioning" of human operators by machines! This very industrial representation of the notion of collaborative work, although interesting, remains very limited (especially from the operator's point of view).

2 IBM engineers admit that they are not able to explain the genesis of the answers developed by their *Jeopardy* champion, Watson!

The announced principle of the "Man–Machine Teaming" PEA is to provide the various machine systems with more autonomy and artificial intelligence for an extended and rethought human–machine relationship. From this perspective, these intelligent systems would no longer be limited to the simple execution of actions requested by an operator. They would allow collaborative work that would make operators' actions and decisions more efficient and effective while saving their mental and physical resources. To this end, these systems would be equipped with increased situational awareness, in particular through different means of perception and analysis (operator status, interactions, prediction of actors' intentions, tactical combat situations, etc.). This capacity would allow systems to learn from the situations encountered, adapt accordingly and share relevant information to support operator decision-making and planning. To guarantee a high level of performance, a guarantee of mission success, this cognitive air system would also integrate new modes of interaction that are more natural and adapted to the situations encountered by operators.

**Box 6.2.** *The example of the "Man–Machine Teaming" upstream study project (PEA), taken from the Dassault Systèmes site*

### 6.4.3. *A mistrust of weak AI*

In a pattern recognition task, machine learning techniques detect correlations between the presence of signs in the same image or scene. A correlation, i.e. the fact that the presence of a given sign is associated with that of another specific sign, does not provide any information on the causes of the joint presence of the two signs. Depending on the field of application, the translation of correlations into causalities can be very damaging since it very often leads to erroneous conclusions. A decision-maker must be able to interpret and validate correlations by crossing other sources of information. In a command structure, AI makes it possible to process large amounts of data (tactical, operational and strategic information) and merge them to propose working hypotheses to decision-makers. We are talking here about a "collaboration" between a decision-maker is a weak AI which, combined with big data, does not solely aim to compensate for certain human deficiencies (low computational and memory capacities). However, decision-makers, aware of the explanatory limits of the proposals developed by support systems, continue to base their decisions on their intuitions and the proposals made by their human collaborators [FER 17]. Let us remember at this level that only human beings are able to connect heterogeneous elements, to learn and understand from little information, whereas weak AI needs a huge amount of data before "understanding" things.

An AI-based system can process a multitude of environmental information in real time: the number and nature of threats, weather, air traffic density, the position of enemies and allies, etc. In the event of an unexpected situation, the decision rests with the command, which must have a perfect understanding of the current situation. This type of collaboration is only possible if the decision-maker is fully aware of the system's capabilities and if the system ensures "behavioral coherence", i.e. provides working assumptions that the command can understand: on the basis of which these assumptions have been developed, are they realistic, feasible, effective? Here, we find the same questions that pilots asked themselves about on-board support systems. However, although the operation of support systems is still understandable to engineers or pilots, this is no longer necessarily the case for AI-based systems.

### 6.4.4. *Strong AI for human–system collaboration?*

Given the technology inherent in weak AI systems and the low interest shown by programmers in human cognitive functioning, the primary purpose of a weak AI system is not to participate in collaborative work with a human counterpart. On the contrary, the objective of the so-called "strong" AI is to reproduce the thinking and intelligent interaction skills (analyze, understand, decide) of humans. Strong AI could solve complex problems, in any environment, with a level greater than or equal to human intelligence all in an autonomous way. In concrete terms, research in strong AI is currently focused on issues of automating different forms of human reasoning and decision-making. In this context, a collaboration that one would describe as intelligent between a human agent and an artificial agent is, in theory, quite conceivable.

Intelligent collaboration is defined here as the implementation of a common activity between a human agent and an artificial agent based on a detailed and reciprocal knowledge of the specificities of both agents (an initial principle identical to that proposed in the sharing of human–machine authority). For a support system to work intelligently with a decision-maker, it is necessary, by adopting a human factors engineering approach, to apply theoretical and operational knowledge about decision-making (collected in particular from the theoretical framework of cognitive sciences) to its deployment. For example, it is necessary to be able to understand human decisions, by identifying heuristics, or biases likely to generate an unexpected, or even inappropriate decision, to translate it into algorithms (i.e. into "frightening algorithms") so that the support system can identify a bad decision (in progress) and propose an alternative, this time, likely to be understood by the decision-maker. To ensure the efficient implementation of autonomous systems, both agents (human and artificial) must consider each other as two collaborators who know each other well, rather than as two competencies that act independently. In the long term, strong AI should be able to reason like a human and thus adapt to the uncertainty of the world while demonstrating almost infinite computational and storage capacities.

## 6.5. Seeing beyond cognition to innovate

Although cognitive psychologists encourage and support AI researchers to consider high-level processes (decision-making, reasoning, language), human intelligence does not reside solely in these cognitive processes. In a recent report on decision-making and the digitization of the battlefield, many Air Force decision-makers still confess to using their intuition to make decisions, even very engaging ones [FER 17]. To offer decision support systems based on strong artificial intelligence technologies, programmers will need to understand what intuitive decision-making is and how intuition "cohabits" with analytical decision-making in the human cognitive system. To hope to innovate in the field of artificial decision-making, specialists in strong AI would undoubtedly benefit from drawing inspiration from Kanheman's [KAN 11] work in which these two concepts coexist (system 1 of intuitive thinking and system 2 of analytical thinking). In addition, there is also an "embodied" form of cognition [TRN 13, VAR 91] in which sensory-motor skills and emotional processes participate in the expression of intelligence [BER 03, DAM 95]. If specialists in strong AI aim to equip their systems with a "consciousness", an ability to feel emotions and/or to understand their own reasoning (a kind of artificial metacognition), they will have to collaborate with specialists in human behavior in a multidisciplinary approach to produce robust and innovative models.

Another major challenge of human–system collaboration, probably the most important and yet the least publicized, is to enable operators to understand what "intelligent" systems do. This is the role of interface ergonomics, or rather, the ergonomics of human–system interactions. Its objective is to make visible, simple and usable what is complex and buried under several layers of processing.

Although a little dated, Hancock and Chignell's [HAN 89] definition of so-called "intelligent" interfaces remains simple and precise. The authors describe as intelligent the interfaces that provide the tools to reduce the representational distance between the mental model that the user of the task has to the way the task is presented to the user by the system. The field of intelligent interfaces covers a disciplinary field representing the intersection between human–system interaction, software ergonomics, cognitive sciences and AI, including their respective sub-disciplines, such as computer-assisted vision, automatic language processing, knowledge representation and reasoning, machine learning, knowledge discovery, planning, artificial and human agent modeling, and language modeling [MAR 14]. Like AI specialists, ergonomists will probably have to imagine interfaces that go beyond the visual modality by drawing on all human resources. A recent thesis [LAR 15] presents a fairly exhaustive set of alternatives to the "all-cognitive" to develop innovative interfaces.

Finally, a significant effort will have to be devoted to the development of specific training on human–system collaboration. Users will have to learn to work with an artificial intelligence system that is superior to them in many tasks while still being, even today, ignorant of the fundamentals of human intelligence.

## 6.6. Conclusion

Strong artificial intelligence will be able to reason like a human being by understanding his experience, habits, preferences, errors, while demonstrating computational and mnemonic capacities beyond comparison with human. Strong AI will be able to correct the human errors of its allies and anticipate those of its enemies. This can only be achieved through an interdisciplinary approach. It is necessary to bring together all specialists in AI: computer scientists, cognitivists, philosophers of the mind, ergonomists of human–system interactions to overcome the technical problems of machine learning and be able to innovate in a field of application very constrained by profitability (a large part of AI specialists work for the GAFA).

Despite the omnipresence of weak AI and the promises of the strong AI, no autonomous system is today able to grasp with as much ease, and ultimately efficiency, the dynamism and uncertainty of the real world as the human being. It is the complexity of the environment that makes human a central part of human–system collaboration. Thus, there is still a long way to go before we can fully understand human beings and in particular the role that emotions play in their decision-making and reasoning. Once this last obstacle has been overcome by a strong AI system that has become fully autonomous, will the question of human–system collaboration still have any meaning?

## 6.7. References

[BER 03] BERTHOZ A., *La décision*, Odile Jacob, Paris, 2003.

[DAM 95] DAMASIO A.R., *L'Erreur de Descartes*, Odile Jacob, Paris, 1995.

[FER 17] FERRARI V., Prise de décision et numérisation de l'espace de bataille: l'exemple du C2. (38 p.), Rapport au profit de l'état-major de l'armée de l'air, 2017.

[FER 06] FERRARI V., DIDIERJEAN A., MARMECHE E., "Dynamic perception in chess", *Quarterly Journal of Experimental Psychology*, vol. 59, pp. 397–410, 2006.

[FER 08] FERRARI V., DIDIERJEAN A., MARMECHE E., "Effect of expertise acquisition on strategic perception: The example of chess", *Quarterly Journal of Experimental Psychology*, vol. 61, no. 8, pp. 1265–1280, 2008.

[GOB 00] GOBET F., SIMON H.A., "Five seconds or sixty? Presentation time in expert memory", *Cognitive Science*, vol. 24, pp. 651–682, 2000.

[HAN 89] HANCOCK P.A., CHIGNELL M.H., *Intelligent Interfaces: Theory, Research and Design*, North-Holland, Amsterdam, 1989.

[KAH 12] KAHNEMAN D., *Système 1 Système 2. Les deux vitesses de la pensée*, Flammarion, 2012.

[LAR 15] LARGE A.-C., Embodied cognition, ergonomie et interfaces de pilotage: le cas des systèmes de drones, PhD Thesis, Université Paul Valéry, Montpellier 3, 2015.

[MAR 14] MARQUIS P., PAPINI O., PRADE H., *Panorama de l'Intelligence Artificielle - Ses bases méthodologiques, ses développements* - Volume 3, L'intelligence artificielle: frontières et applications, Cépaduès, 2014.

[MUC 17] MUCCHIELLI R., *La dynamique des groupes*, Editions ESF, 2017.

[PEL 05] PELLETIER F.J., ELIO R., "The case of psychologism in default and inheritance reasoning", *Synthesis*, vol. 146, pp. 7–35, 2005.

[PIE 03] PIETARINEN A.-V., "What do epistemic logic and cognitive science have to do with each other?", *Cognitive Systems Research*, vol. 32, pp. 169–190, 2003.

[SIM 96] SIMON H., *Les sciences de l'artificiel*, vol. 3, Gallimard, 1996.

[SIM 97] SIMON H., *Models of Bounded Rationality*, vol. 3, MIT Press, 1997.

[TRN 13] TRNINIC D., ABRAHAMSON D., "Embodied interaction as designed mediation of conceptual performance", in MARTINOVIC D., FREIMAN V., KARADAG Z. (eds), *Visual Mathematics and Cyberlearning*, pp. 119–139, Springer, New York, 2013.

[TUR 50] TURING A., *Computing Machinery and Intelligence*, Oxford University Press, Mind, vol. 59, no. 236, 1950.

[VAN 08] VAN BENTHEM J., "Logic and reasoning: Do the facts matter?", *Studia Logica*, vol. 88, pp. 67–84, 2008.

[VAR 91] VARELA F.J., THOMPSON E., ROSCH E., "The embodied mind", *Cognitive Science and Human Experience*, MIT Press, Cambridge, MA, USA, 1991.

# 7

# Perspectives and Ambitions of the Maintenance in Operational Condition Renovated at the Heart of the Armament Programs: Illustrations in the Terrestrial Environment

ABSTRACT. The MCO is marked both by the multitude of actors involved and by a necessary evolution of the structures to face the new challenges of the Defence. Innovation is an important focus of the MCO in terms of maintaining or increasing operational technical availability while controlling costs, throughout the life cycle and in the framework of participation in international cooperation.

## 7.1. Introduction

The maintenance in operational condition (MCO) is marked both by the multitude of actors involved (Military Staff, DGA, users, support services, private or public industrialists, etc.) and by the necessary structural developments required to meet new challenges in the defence world. Innovation is an important focus of the MCO in order to maintain or increase operational technical availability (DTO, i.e. the ability to conduct operations) while controlling costs throughout the life cycle. Certain normative requirements (e.g. REACH) also make MCO actors adapt. Innovation can also be part of international cooperation opportunities.

---

Chapter written by Nicolas HUÉ, Walter ARNAUD and Christophe GRANDEMANGE.

From the perspective of the Directorate General of Armaments/Direction Générale de l'Armement (DGA), the formation of the MCO is developed during the successive stages of an Armament Program; we will discuss its challenges and perspectives.

In the first part of this chapter, we present the context and challenges of innovations for the MCO. In the second part, we then analyze the advantage of digitization for the OLS, before examining prospects for improving OLS performance through digital innovation in the third part. We conclude our discussion with field observations on future innovations.

## 7.2. Context and future challenges of the MCO

### 7.2.1. *End-to-end construction, from upstream phases to the in-service use phase*

In terms of OLS, the DGA's involvement is decisive in the upstream phases, in the implementation and qualification phase of the system and in the initial support phase. The DGA also remains heavily involved in the in-service support phase. More precisely, the DGA defines the support system and associated performances according to staff requests, evaluates the overall cost, contracts the acquisition and the initial support phase, transfers the system to the forces after qualification, conducts system upgrades and maintains technical training throughout the life of the program.

– In the *upstream phases*, the development of a support strategy becomes a key tool to help define the main options.

More specifically, this support strategy first helps analyze the various possible technical solutions and identify the preferred one. Second, the distribution between industrial and operational work is analyzed: the existing organization is given feedback, solutions from similar programs and the implementation of innovative results that are more effective in terms of overall cost, efficiency and the meeting of performance objectives. Next, the main contractual options are presented, in order to plan the following phases. Finally, an inventory of innovations and their possible implementation is presented.

This support strategy, which is validated by all project team participants, helps to structure the possible options and identify the preferred option(s).

– In the *implementation phase*, the challenge is to make a long-term effort and to engage manufacturers so that they can organize themselves and demonstrate they meet the targeted performance levels in the long-term when the project is launched and with detailed designs to choose. The contractual formulas including acquisition, initial support and part of the in-service support are then those that provide the most guarantees and facilitate the handover between the prime contractor, the DGA and the support services.

– In the *use phase*, rapid technological change requires regular system upgrades and requires consistency between these upgrades and supporting activities. These upgrades can involve significant revamps and therefore have a significant impact on availability. In this case, the contractual formulas including both building sites and in-service supporting activities on the rest of the fleet allow greater control over availability.

All of these actions contribute to the end-to-end organization of MCO's activities. The involvement of the DGA will then gradually decrease once the upgrades are completed, with the DGA maintaining technical control until the dismantling phase.

### 7.2.2. The necessary awareness of stakeholder responsibilities

Faced with these challenges, a so-called *per environment* approach is being implemented within the Military, entrusting a "Lead Chief of Staff" with responsibility over the performance of the MCO serving land and naval environments (Chief of Staff of the French Army (CEMAT) and Chief of Staff of the French Navy (CEMM), respectively). Regarding the management of the MCO, this approach focuses on both the annual definition of performance objectives and the refocusing of responsibilities according to environment. Through the MCO aeronautical transformation plan, the management of the aeronautical environment, which includes a strong joint dimension through its fleets of helicopters, UAVs and fighter aircraft, is entrusted to the Armed Forces Headquarters, with the establishment of the French Aeronautical Maintenance Directorate.

On an industrial level (private or public), efficiency requires greater contractual responsibility for a support service based on engaging objectives that make it possible to achieve availability objectives in controlled budgets.

### 7.2.3 *What are the support mechanisms for better industrial accountability?*

The answer varies from one Program to another and is often customized according to operational needs and constraints, as well as to the assessment of product maturity, and therefore the level of risk to be transferred to the manufacturer.

The *Global Service* strategy allows for a broad accountability of the manufacturer. This is the example of the FOMEDEC training program (Modernized Training and Differentiated Training for Fighter Crews), which provides for the use of new training aircraft (Pilatus PC 21) and associated simulators for a period of 11 years.

The *Fleet availability* strategy is chosen if the objective is to obtain a commitment by the hour of flight, for a duration of mission or for a number of kilometers covered. This is the example of the Unprotected Versatile Tactical Light Vehicle Program (VLTP-NP), with a market combining the acquisition of 3,700 vehicles and 20-year support based on a minimum 95% availability requirement for the fleet (excluding OPEX).

Finally, a *Parts Availability* strategy (under the single window type of approach) is also used to give the manufacturer complete responsibility over parts management, with a commitment to a delivery time. This is the case for in-service support for the M88 engine of the Rafale aircraft, with the implementation of an engine module window and a commitment from the industry on the availability of these modules.

These strategies are thus preferred over a simple commitment to revamp deadlines, which provides only a partial guarantee of the result obtained and fragments responsibilities.

### 7.2.4. *The influence of the environment*

In order to optimize the chosen solution, this support service must also take into account the benefits that the related environment can bring.

*Cooperation* offers opportunities for operational and financial gains that can be significant: sharing of technical monitoring, joint configuration management, multilateral pooling, scale factor, smoothing of peak loads, training, documentation, joint maintenance, etc. This is, for example, what the A400M Program is all about: setting up a shared support system, initially formed between France, Great Britain and Spain. The model can go as far as the implementation of more global solutions covering all levels of intervention, with the exception of national operations on bases, which are generally dedicated to each nation.

The establishment of the EMAR (European Military Airworthiness Requirements) regulation in the air transport sector facilitates the mutual recognition of maintenance operations between States.

Finally, civil–military *duality* offers opportunities that should be grasped, for example, a park derived from a civil park. Military support is then based on a civilian rear base generating a scale factor and allowing the use of existing structures. This is the example of the MCOs set up, for example, for lightly armed buildings, for certain vehicles (excluding armored vehicles) or for military aircraft derived from commercial aircraft, most of whose equipment comes from the civilian sector (Falcon 50 maritime surveillance, tankers, etc.).

Finally, with the multiplicity of interventions by armies outside metropolitan France (OPEX), support for OPEX is now an integral part of the support services. The difficulties relating to logistical flows between metropolitan France and the theater of operations, which are essential to achieving the expected performance, may require new solutions: a window as close as possible to the forces deployed when the theater of operations allows, improved routing, including customs clearance times in host countries, or even the possibility of setting up repair workshops directly in the theater of operations.

### 7.2.5. *Financial issues that are central to the work*

In the face of budget pressures, optimization must be sought. By usually considering that, for a given program, the total cost of the OLS is more than twice the cost of its implementation phase, we can measure the associated economic stakes.

For the DGA, this means taking into account as soon as possible the estimated overall cost, including the amount of industrial contracts and all other costs internal to the Ministry (wages and social security contributions). This overall cost is presented by the DGA for the launch of the various stages of the Program and is a common tool used by all stakeholders.

With regard to the formation of future French budget legislation, detailed analysis on a fleet-by-fleet basis, or better still, on a program-by-program basis, is crucial in order to check physical and financial consistency and set priorities. In more detailed financial management, an objective cost per program, fixed at the outset and whose changes are monitored for the duration of the program, would be useful to make links between the implementation and control strategy from start to finish and in-service support costs.

## 7.3. Innovations for the MCO of the future: the prerequisite for digitization

### 7.3.1. *The necessary digitization of the MCO*

The real breakthrough that will be brought about by the innovative actions led by the Directorate General of Armaments (DGA), in full agreement with the French Armed Forces Staff (EMA), is to proceed with the digitization of the MCO. In fact, as soon as the MCO data are – finally – digitized, all information technologies can become applicable to the MCO: information and management systems, constrained optimization algorithms (availability, budget, HR, etc.), virtual or augmented reality, artificial intelligence (in particular predictive algorithms to anticipate heavy actions), big data, etc. The current use of paper format is a hindrance to the development of innovations to the benefit of the MCO, while not guaranteeing a sufficient level of performance.

However, this objective is not purely technological. The aim is to improve the availability and knowledge of the reliability of current or future platforms, reduce maintenance costs, increase safety when using systems or reduce the maintenance burden at the operational level.

In this context, the DGA, the EMA, the general military staff (land, air, sea) and the support structures have agreed on "innovation roadmaps", with overall coherence and preserving the appropriate specificities according to the environment (land, air, sea, IS, ammunition). These roadmaps provide plans spanning several years (5–10 years), with an overall consistency and – above all – a vision shared by all state stakeholders, which makes it possible to give real visibility to industrialists.

It should be noted that while these roadmaps are state plans, they naturally take into account the state of the art in industry and known high-potential innovations. They also take into account the knowledge of foreign armed forces and the innovations they lead for their MCO.

### 7.3.2. *The foundation of digitization: RFID, HUMS and interoperability*

#### 7.3.2.1. *Radio-frequency identification (RFID)*

The DGA and the SIMMT, the French Defence Ministry's integrated through-life support structure for land forces equipment and systems, have resolutely oriented themselves towards the deployment of RFID technology for the traceability of materials and the optimization of the logistics management of goods. The EMA and the DGA are preparing a joint policy on RFID tag deployments, which will be applicable to the various environments.

Two standards, meeting international standards, have been adopted: HF (ISO 15693 standard) and UHF (ISO 18000-63 standard, GEN 1 or GEN 2 encrypted generation).

RFID is a technology that is already well-known and objectively supported by the civilian world. Nevertheless, military applications required several precautions and adaptations.

First of all, it was essential to ensure that RFID tags, which are antennas and therefore radiating elements, did not present any risks in the face of improvised explosive devices (IEDs). Experiments were thus carried out by the STAT (Technical Section of the French Army) and radiation measurements were taken by the DGA/TA (Advanced Techniques) test center. In addition, RFID tags and their environment (readers, SIL, etc.) are also SSI certified, so that this technology is not an entry point for a computer attack to paralyze the supply chain.

The HF standard radiates at less than 10 cm and is intended to be used on infantryman equipment, insofar as this equipment is stored in armories. In fact, the presence of the RFID tag is not enough: it must be coupled with a "physical" control by the personnel (it is not a question of knowing that 100 RFID tags are in the armory, but a question of being sure that the expected weapons are in fact present!).

The UHF standard is intended to be used on vehicles, but positioned on-board the vehicle minimizing any radiation to the outside. The choices will therefore be defined during joint test plans, defined between the DGA and the French Armed Forces.

The use of RFID tags has already demonstrated substantial advantages: verification operations that used to take 48 hours can now be done in less than an hour. The extracted data are free of handwritten transcript errors and can be considered reliable. The logistics management of assets is more efficient, and staff can be assigned to tasks of higher added value.

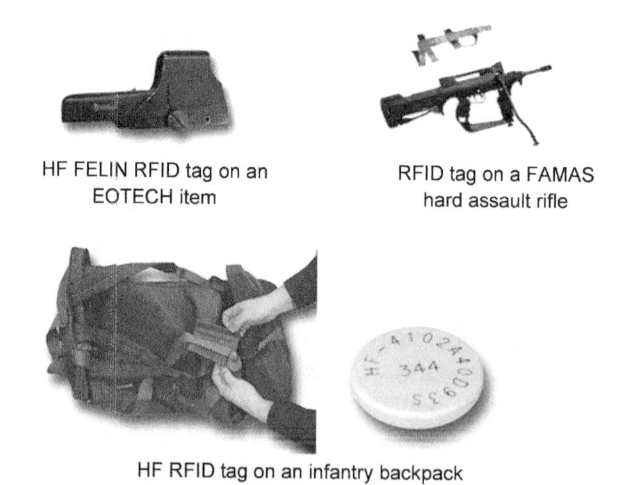

HF FELIN RFID tag on an EOTECH item

RFID tag on a FAMAS hard assault rifle

HF RFID tag on an infantry backpack

**Figure 7.1.** *Example of the use of RFID tags. For a color version of the figures in this chapter see www.iste.co.uk/barbaroux/technology.zip*

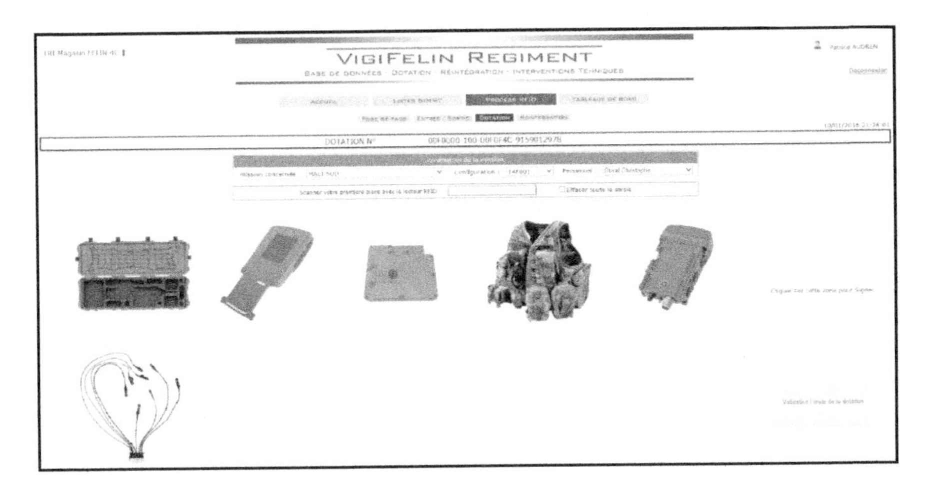


Figure 7.2. *Window of the VIGIFELIN portal*

For vehicles, RFID tags are inserted inside vehicles, for example:

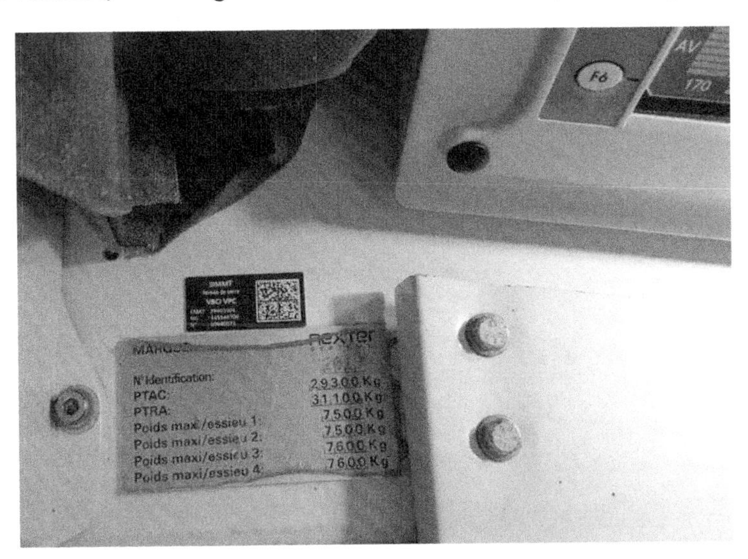


Figure 7.3. *UHF RFID tag inserted inside a VBCI armored vehicle*

Finally, in order to digitize the MCO, data relating to equipment can be integrated into information systems and allow for fine control. For example, in the case of FELIN, RFID tags make it possible to know in near real time the state of a company's equipment: in the event of a projection, the land Forces command can thus know which regiment, and even which company/squadron is able to be immediately projected in relation to its available equipment.

### 7.3.2.2. *HUMS (Health and Usage Monitoring System)*

In the land sector, the DGA, the Staff of the French Army (EMAT) and the SIMMT (Integrated Structure for the Operational Maintenance of Land Equipment) have developed a roadmap for the deployment of HUMS (Health and Usage Monitoring Systems) sensors in the French Army, in compliance with operational safety and ISS rules, and designed to provide dematerialized information on the state of fleet health. An experiment was carried out as part of a Technical-Operational Study (TOS) (called "PROPHETE"), which was completed on March 14, 2017. The results were considered to be promising, this TOS will be supplemented by a larger experiment in 2017–2018 (on a fleet representative of a unit in combat): an evaluation in operational context (EVTA) will be conducted by EMAT on a fleet of 20 VAB armored vehicles in order to clearly identify the needs of maintenance personnel, the ergonomics and the volumes of the data flows to be stored and/or transmitted.

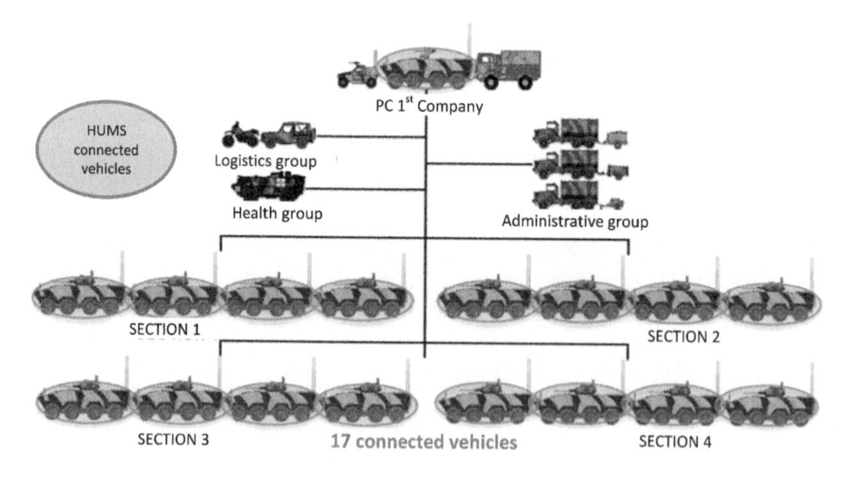


**Figure 7.4.** *Health and Usage Monitoring Systems: case of a combat company using connected vehicles*

Several requirements have been clarified for HUMS: they must be modular (installation/uninstallation in less than half a day), as generic as possible (battery HUMS, fuel HUMS, etc. whatever the vehicle), and SSI approved. In addition, the data produced must have a non-proprietary format (text format) and will be the property of the State.

The HUMS must of course allow the SCORPION program to benefit from predictive maintenance, i.e. maintenance that can be managed and planned.

However, HUMS are not intended to be limited to the SCORPION program alone. Other vehicle fleets may be eligible, provided that the benefit is proven in relation to the installation costs. Thus, HUMS are already planned on several land equipment: assault rifle (shot counter), laser designator (health of the laser cavity), missile (MMP), etc.

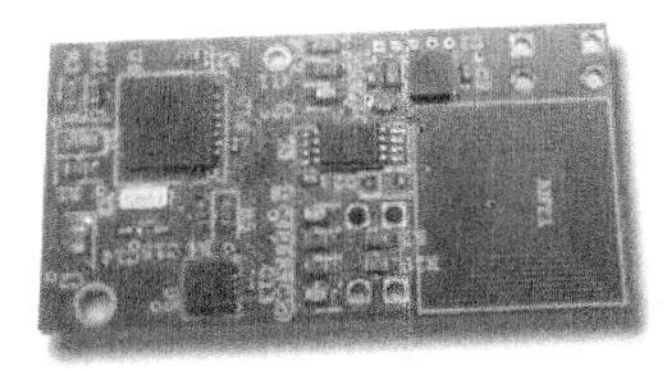

**Figure 7.5.** *HUMS FAMAS stroke counter – DGA/DT/MIP and SIMMT development*

In this context, the Integrated Ammunition Service (SIMu) has also engaged with the DGA on an "innovation roadmap" providing for the deployment of HUMS for ammunition, and a gradual transition to planned maintenance. The main challenge with ammunition is data recovery. Indeed, wireless technologies, which are the most practical, must be qualified to protect against any coupling effect that could lead to a pyrotechnic risk.

### 7.3.2.3. *Interoperability of information systems for data sharing*

The data produced by the RFID for traceability and by the HUMS for the "health status" of the equipment will be implemented in information systems (logistics). To exploit possible correlations, global feedback, job profiles, etc. it is essential that information systems can be exploited as a whole, and are therefore interoperable.

This interoperability is all the more essential given that the digitization of MCO activities (ordering parts, monitoring flows, etc.) must be ensured from start to finish, without any disruption that could potentially lead to loss of data, integrity or reliability. With regard to these information systems, the DGA, EMA and the French Army have therefore adopted the PLCS (Product Life Cycle Support) standard to enable the interface and interconnection of logistics information systems, both national and industrial (as well as with NATO).

For SMEs, lighter versions of the standard are planned so as not to penalize them. Nevertheless, the PLCS standard will ensure – for all IS – the interoperability and consistency of data.

## 7.4. Innovations for the MCO of the future: research and innovation challenges

### 7.4.1. *Predictions of optimal maintenance plans: artificial intelligence and big data*

Assuming that the OLS will eventually be digitized globally, the data produced – supposedly exhaustive, reliable, honest – contained in the LIS can be used by all the innovative methods available.

The term predictive maintenance can be misinterpreted. It is not a question of imagining maintenance acts. Rather, it is a question of being able to manage and plan these acts, by fine-tuning the state of health of the equipment. There is therefore, very logically, an objective of predicting maintenance actions based on the evolution of a life profile. It therefore appears that artificial intelligence algorithms, in particular predictive algorithms, and big data will be disciplines with high added value for the MCO.

Despite the connotation of “big”, it is not necessarily useful to have huge volumes of data. Temporal depth and accurate (reliable) data may be sufficient to build predictive models. The French National Center for Space Studies (CNES) has already undertaken research in this direction and applications at the military MCO seem very appropriate. The first step is therefore to quickly and proactively develop digitization through the deployment of HUMS and RFID tags to be in the short-term capacity to implement optimized maintenance plans in the military field.

### 7.4.2. *Augmented and virtual reality (AR/VR)*

The main difficulty is that AA/VA must be consistent with the known configuration of the equipment to be maintained (applied theoretical configuration). It is therefore essential that information systems (IS) have access to real-time equipment configuration information. Indeed, AR/VR could be used to measure the differences between the theoretical and actual configuration.

In addition, AR/VR will make it possible to simulate the best arrangements of a vehicle to optimize maintenance operations (engine removal, emptying, etc.) according to the parameters to be optimized (cost, delays, cost/time ratio, number of personnel required, etc.). For example, AR/VR already makes it possible to simulate the position of tools in a workshop to optimize acts, time per act, etc. while ensuring that workplace safety conditions are respected.

AR/VR can also be usefully applied to training on maintenance procedures, subject to consistency with the configuration of the equipment to be supported.

### 7.4.3. *3D printing*

An application results from the configuration constraints necessary for 3D printing. If the exact configuration – theoretical and/or real – of the equipment and its parts is known, then 3D printing of certain parts from the digital configuration files becomes perfectly feasible.

SMCO is currently studying the range of parts and equipment that could be suitable for 3D printing in the long term: limited complexity, high performing, regular wear, highly used consumable parts, etc. Indeed, the knowledge and updating of a digital configuration file is costly and it seems appropriate to properly define the appropriate requirements. It is also necessary to define the contractual clauses that will allow the gradual transfer from one industrial company to another (in particular, state) to continue the 3D printing of parts.

### 7.4.4. *Remote maintenance*

Finally, still as part of the technological innovations for the MCO, the SMCO service of the DGA evaluates remote assistance/remote maintenance capabilities. Indeed, if OLS data are digitized and dematerialized, it is possible to draw inspiration from remote assistance precedents, particularly remote medicine: for subjects with complex diagnoses (requiring rare expertise) but with simple solutions, remote assistance makes it possible to achieve excellent results. This is the case for strokes in medicine, where the diagnosis (neurology) is very complex, but the treatment is relatively simple (feasible by an emergency doctor). Having digitized the OLS, with the SIL and associated transmission methods, remote maintenance will make sense. For these complex cases, remotely interviewing an expert in metropolitan France instead of planning (if at all possible) will save valuable time and money. Applications have already been successfully applied to equipment such as the LECLERC tank, the armored infantry fighting vehicle (VBCI) or even the unit rocket launcher (LRU). This goal still needs to be developed and outlined.

## 7.5. Some safeguards

### 7.5.1. *Technology at the service of humans*

HUMS, RFID tags, etc. are valuable tools. But they remain tools. Establishing a maintenance plan is an onerous and complicated operation, requiring the implementation of infrastructures, human resources and contractual means. Even though the idea of an MCO “agility” is attractive, it will be limited in reality: industrialists and operations cannot spend their time developing maintenance protocols, despite its priority being the success of military operations, particularly in theaters of operations. The peace-crisis-war continuum of the EMA remains and must remain an essential condition to be respected: regular adaptation is appropriate; permanent disorganization is not acceptable.

Similarly, AR/VR, remote maintenance, etc. will help operators and maintenance staff. But under no circumstances will these personnel lose their own responsibilities.

In short, these technological innovations will enable Forces to understand, manage and control their OLS, and to adapt it when necessary based on objective criteria. However, decision-making remains and will remain a human task.

### 7.5.2. *Jobs and skills that need to be managed in symbiosis*

The support services, which have been created since 2010, need to carry out their mission in professions similar to those at the heart of the DGA's missions: managers, architects, purchasers of complex acts, technical experts, program quality experts, financial managers, etc. A common job and skills planning management system (GPEC) is therefore being set up to manage the sectors and strengthen the professionalism of staff.

With regard to the pace of operations, it must be objectively considered that these innovations will save time, achieve better DTO, and keep costs and resources in check. To consider that they would be a factor in saving staff would be very hypothetical.

### 7.5.3. *A strategic challenge for the DITB*

The implementation of innovations is also a guarantee of competitiveness, strengthening the export position of industrialists to the benefit of the defence industrial and technological base (DITB). The offer of support must be competitive and adapted to new concepts.

The digitization of the MCO will allow better planning, as well as better control of costs/delays/optimization of production chains. For the industry, this guarantees a much better visibility of load plans, and even control of these.

For the preservation of the OLM DTIB, this is a particularly important asset, especially since the OLM solutions used by French Forces can also be proposed for export.

## 7.6. Prospects for the future

In short, all the facets of the DGA expertise are in the process of setting up the MCO, in fact, similar to what is being done for acquisition guidance: definition of systems, contractualization, involvement of the various trades, innovation, etc.

The main areas of effort mainly include an end-to-end approach fueled by a strategic vision, greater accountability of stakeholders and greater control over financial issues.

Finally, regarding long-term external operations, the resilience of systems is put to the test. The conditions of use of the equipment (dust, sand, temperature, etc.) lead to increased constraints and a redefinition of support formulas in order to use the theaters of operation requirements to define the support, to optimize logistics flows and to take into account feedback from operational experience. The field of operational maintenance is thus facing new challenges.

# 8

# Technological Change and Individual Competencies: The Influence of Glass-cockpit Aircraft on French Air Force Pilots Training and Skills

ABSTRACT. This chapter investigates the influence of introducing new generation aircraft on the training of Air Force pilots. New generation aircraft (Cirrus SR20 and Pilatus PC-21), equipped with novel capabilities (glass cockpit and embedded training), represent a technological disruption that is likely to affect the nature and diversity of pilots' competencies, and more globally the learning process that leads to their development. Building on past research projects that the authors conducted for the Air Force Staff and Pilot Training Schools, this chapter first shows that the new functionalities offered by next generation aircraft are not neutral on how future pilots develop fundamental technical skills ("flight basics") during the initial phases of the training process. Second, it identifies the conditions enabling a consistent exploitation of embedded simulation techniques for training pilots.

## 8.1. Introduction

Technology and skills are closely related [GOL 98]. Many studies identify the existence of a close relationship between the properties of the technologies used to produce, manage and communicate information or assist in decision-making, and the skills of the individuals responsible for their design, maintenance and use [ADL 91; BRE 05; GAL 02]. If we consider, with [CAP 93, p. 516], that "*technology determines the nature of*

Chapter written by Cyril CAMACHON and Pierre BARBAROUX.

*the necessary skills*", then any modification affecting the socio-material properties and/or functionalities of the technology essentially leads to a reconfiguration of individuals' skills or even their qualifications [LEI 96; OIR 05].

Aerospace and defence (A&D) organizations are particularly sensitive to technological innovations, in particular those concerning aircraft capabilities [BAR 11] and command and control systems ($C^2$, [GOD 10]). The digitization of cockpits and $C^2$ environments, the automation of certain pilotage, navigation or air traffic control tasks, and the hyper-connectivity of aircraft and operators are new technological capabilities to which individuals, crews and organizations must adapt.

In this context, we question the effect of the changes brought about by the introduction of new aircraft on the skills of French Air Force pilots. In particular, we examine the implications of the use of the Cirrus SR20 and Pilatus PC-21 aircraft for pilot training and skills development, two aircraft equipped with modern glass-cockpit avionics[1] and, for PC-21 only, an on-board simulation capability. Within this framework, we study the influence of the introduction of the SR20 on the process of acquiring basic technical skills ("flight basics") during the initial pilot training phase. The training model for Air Force pilots is based on a breakdown of the training process into phases, the objective being to support the development of the trainee pilot's expertise through the progressive acquisition of three categories of skills: technical, relational and situational [BAR 12]. Thus, at the end of the initial training phase (École de l'Air in Salon-de-Provence), the trainee pilot must demonstrate mastery of a basic set of technical skills relating to the performance of elementary maneuvers applied to piloting (taxiing, take-off, navigation, landing, basic aerobatics), through the mobilization of a set of behavioral attitudes in accordance with the principles guaranteeing flight safety (visual scanpath, flight parameter control loop, navigation and situation awareness; [DUB 15]). This crucial phase of the training process is also carried out within a constrained timeframe (27–36 weeks maximum), the acquisition of elementary skills ("stick-and-rudder skills") based on the

1 Modern cockpits are composed of two LCD screens replacing the traditional dials: the PFD (primary flight display) allows the display of primary flight information and the MFD (multifunction display) is used to access the various management pages of embedded systems (GPS, background map, flight plan, etc.).

repetition of a cycle composed of three (flight) sequences: discovery, implementation and validation.

Basically, the introduction of aircraft equipped with glass cockpits has an impact on pilot training and associated skills which can be analyzed on two levels. First, it requires the mastery of body language ("buttonite") adapted to the use of a digital interface; it also involves the development of new skills relating to the management of the information space, these new skills affecting by extension the dynamics of building the pilot's situational awareness. Second, the introduction of Cirrus SR20 (glass cockpit, initial phase) and Pilatus PC-21 (on-board simulation) aircraft opens up a field of questions relating to the decomposition and distribution of training content during the pilot's training path. These two levels of analysis appear relevant to our objective.

The rest of this chapter is organized as follows. Section 8.2 presents the pilot training model and techniques as well as the typology of the skills developed. Section 3 introduces the context of the research (data collection and analysis, presentation of the research context). Section 4 reports on the developments observed in the acquisition of basic technical skills ("flight basics") during the initial training phase on Cirrus SR20. It also discusses the implications of these developments with regard to the issues raised by the use of on-board simulation techniques in the subsequent phases of the pilot training program. Section 5 elaborates on the implications of the research and draws a conclusion.

## 8.2. The pilot training model: epistemological foundations and typology of skills

Pilot training techniques provide individuals and crews with critical resources to develop experience within both real and simulated flight contexts. Therein, learning results from the accumulation of experiential knowledge, the latter being "*both the subject and the product of learning*" [BAR 16, p. 42].

The French Air Force has chosen to train its pilots by combining different simulation-based training procedures. These processes are part of the reference framework of LVC simulations – *Live* (Real Operators and Equipment), *Virtual* (Real Operators – Simulated Environments and

Systems) and *Constructive* (Simulated Operators and Systems), or their combination (e.g. embedded simulation). These different processes are particularly attractive in that they multiply the opportunities for accumulating experience and acquiring skills [BOL 09] in environments which, depending on the training objectives, may be realistic or imaginary, simple or complex. LVC simulation thus makes it possible to confront individuals with critical events under controlled conditions (e.g. costs, deadlines, safety) [SAL 98; 09].

The initial pilot training phase is based on a combination of real flight sequences and simulated ground flights. Pilots in training alternate between the discovery, implementation and validation of ground and flight skills, with the consolidation of skills developed according to a knowledge transfer mechanism [NOK 09]. More broadly, BAR [10] and BAR [12] have shown in two case studies on Air Force transport squadrons that the skills acquired and applied by flight crews can be classified into three categories: technical, relational and situational. Technical skills refer to the mobilization of knowledge and expertise related to piloting tasks (landing, take-off, navigation, etc.), the implementation of general and specific aeronautical regulations (checklists, flight safety), and the monitoring of automation (e.g. autopilot management). These skills are developed throughout pilot training courses until the "combat ready" qualification is obtained.

During the initial training phase, technical skills are limited to understanding the aircraft's "physical" behavior ("sense of the air"), performing basic maneuvers related to piloting the aircraft ("flight basics") and rudimentary management of navigation and communication systems (information space management). While their development begins during the initial pilot training phase, relational and situational skills are mainly cultivated in later phases, the aim being to enable the pilot to master the challenges related to the implementation of a weapon system in more complex decision-making environments. In this type of environment, mastery of internal and external communications (e.g. standardization of languages, use of communication systems) and management of collective dynamics (e.g. crew synergy, mission briefing–debriefing, knowledge of codes and work practices in joint and international organizations) become predominant. This is also true for situational skills, which refer to the set of skills related to decision-making (e.g. sensemaking) and managing a complex tactical environment (e.g. research and synthesis of tactical information, real-time management of formation flights, proposing certain maneuvers, etc.). It should be noted that relational and situational skills are not independent and that the associated knowledge types (i.e. know-how, know-what, know-who and know what to do) are developed during the different phases of the pilot training and, more generally, during a pilot's operational career.

The following sections explore the changes brought about by the introduction of Cirrus SR20 and Pilatus PC-21 aircraft on pilot skills and training. We first present the context of the research (data sources and analysis, description of the initial training phase), and then present the results by distinguishing two levels of analysis: changes in the content and methods of acquiring technical skills, and adaptations of the decomposition and distribution of training programs within the different phases of the pilot training program justified by the possibility of exploiting very early on, the embedded simulation capability offered by the Pilatus PC-21.

## 8.3. Research context

### 8.3.1. *Data sources and analyses*

The authors of this chapter collected a wide variety of data during several technical–operational studies for the French Air Force between 2005 and 2017. These studies enabled them to work in contact with officers in charge of the planning activity, command and control of operations (network-centric and $C^2$ operations, 2005), within fighter squadrons (introduction to Rafale and the evolution of skills, 2009), transport squadrons (introduction of the A400M and transfer of skills, 2010), UAV operators (training track, 2010) and pilot training units (ground simulation and embedded simulation, 2012; systems management in initial training, 2014). In addition to the diversity of research topics (e.g. command and control systems and skills, and education and training processes), organizations (in France and abroad) and stakeholders (e.g. senior officers, pilots, navigators, flight engineers, weapon systems, drone operators, trainers, instructors, etc.), these studies have led the authors to explore the multiple dimensions of technological change and its individual and organizational implications.

This contribution is the result of a multidisciplinary research (management sciences and cognitive sciences) that capitalizes on four separate studies, in which the authors participated (see Box 8.1). These studies propose field surveys conducted, for the first three (2009, 2010 and 2012), using a qualitative case study methodology [YIN 05] and, for the last one (2014), in accordance with the experimental method principles. These surveys involved the collection and analysis of primary data (39 semi-directive interviews, participating and non-participating observations during dozens of briefings and debriefings of missions, training and education sessions and exercises in real or simulated flights, experiments on a population of student pilots and flight instructors). Primary sources were supplemented by a variety of secondary data (institutional, conceptual and doctrinal documentation, previous studies, professional publications). According to the studies, the thematic analysis of data was carried out by triangulation and open coding (manual processing or using NVivo software). The robustness of the categories and the coding list was obtained thanks to successive round trips to operational staff.

1) [CAM 14]. Impact of the introduction of systems management (SM) training modules on the learning of flight fundamentals in phase I (Synthesis Note – 6 pages). Research contract on order of the Flight Crew Training Schools (E.F.P.N). This study contributes to studying the impact of systems on flight control, and to defining the added value and pitfalls associated with the handling of these systems in flight, in order to propose recommendations for instructors.

2) [CAM 12]. Ground and embedded simulation (56 pages). Research contract commissioned by the Air Force Planning Bureau (E.M.A.A.). This study explores the limitations and risks of the use of simulation (LVC) in the training and education of flight crews, focusing in particular on the impact on aviation safety of the ability to "inject virtual into real systems" offered by modern training aircraft.

3) [BAR 10]. Maintenance and transfer of skills in tactical transport squadrons: from Transall to A400M (105 pages). Research contract commissioned by the Air Force Planning Bureau (E.M.A.A.). This study explores the dual issue of maintaining the skills of Transall flight crews and transferring them to A400M crews.

4) [BAR 09]. The Air Force's skills by 2015 (125 pages). Research contract commissioned by the Air Force Planning Bureau (E.M.A.A.). This study helps to identify the key skills associated with the use of new complex systems and technologies (Rafale aircraft and Tactical Data Link, Liaison 16, L-16) in Air Force squadrons.

**Box 8.1.** *The four studies on which this contribution is based*

### 8.3.2. *The initial training phase at Salon-de-Provence*

The French Air Force provides *ab initio* training for its aircrew (fighter pilots, transport pilots and combat navigators). For the "Fighter pilot" course, for example, the institution recognizes a pilot as an "expert" once he has obtained the qualification of Flight Leader (FL). The route to this qualification requires an average of eight years of training on school aircraft and weapon aircraft. With regard to the initial "school" period, the training is divided into four main phases. Initial aeronautical training – phase 1 – takes place at Salon-de-Provence on gliders and Cirrus SR20 light aircraft. During this phase, the student pilots essentially learn to master visual flight. During phase 2, carried out in Cognac, they learn to master instrument flying, to practice aerobatics and to carry out more advanced and complex navigations. The "Fighter pilot" pre-specialization leads the selected student pilots to phase 3, in Tours, where they will learn to fly a jet (Alphajet). Finally, phase 4, carried out in Cazaux, should allow the acquisition of elementary tactical skills (also on Alphajet). At the end of these four phases, the student pilots, who have become "trainees", will still have to spend a few more months in specialized squadrons in order to receive specific training on the weapon aircraft on which they will fly when they join their first assignment squadron.

For phase 1 (Salon-de-Provence), the Cirrus SR20 succeeded the Socata TB10 in the summer of 2012, taking a major technological leap forward. Indeed, the cockpit is moving from on-board instrumentation dating from the mid-1970s (multidials, multigauges) to instrumentation composed of digital displays (see Figure 8.1).

**Figure 8.1.** *Analog cockpit (left) versus glass cockpit (right). For a color version of the figures in this chapter see www.iste.co.uk/barbaroux/technology.zip*

The Cirrus SR20 is part of this category of modern aircraft qualified as *Technically Advanced Aircraft* (TAA) due to the presence of modern avionics equipment and systems on board. Indeed, in general, TAAs are aircraft in which the pilot must interact with one or more computers in order to fly, navigate or communicate. According to the US Federal Aviation Administration (FAA), a TAA is an aircraft equipped with *at least an* IFR-certified GPS navigation system, a *Moving Map Display* system and an integrated autopilot. However, many aircraft classified in the TAA category also have a multifunction display (MFD) to display weather conditions, air traffic or topography.

This change in the level of embedded technology is not without consequences when learning to fly [FUN 99]. Thus, when the Cirrus SR20 arrived at Salon-de-Provence, it appeared oversized in the eyes of some phase I instructors, since the objective of this phase is essentially to enable young student pilots to acquire "flight basics". Although there is no official definition, military instructors refer to the concept of flight fundamentals to define the basic skills required to operate a flight without tactical consideration. These are fundamental skills that allow an aircraft to fly safely in the third dimension. The basics can be translated into basic actions from which more complex skills develop. Traditionally, they can be grouped into five main areas (see Table 8.1).

| Domains | Competencies |
|---|---|
| Piloting | Navigation: pitch altitude and inclination<br>Maintaining and respecting the trajectory<br>Altimeter settings<br>Attitude–speed relationship |
| Surveillance of the sky | Level flight<br>Flight while turning |
| Vital actions | Periodic<br>Ascent/descent |
| Navigation | Departure and arrival procedures<br>Visual navigation<br>Blind flight |
| Communication | Phraseology<br>Radio management |

**Table 8.1.** *Basic skills grouped by domains of competence*

The objectives of phase I are to provide student pilots (also called cadets) with the skills necessary to continue their training while assessing their potential for progression. Through different modules, cadets are assessed on a permanent basis, and any individual whose progression is considered too slow can experience the permanent cessation of their pilot training. During this phase, we find, for example, the modules basic navigation (BN), general handling (GH) and basic landing (BL) which contribute to the acquisition of basic piloting and navigation skills. In addition, the basic aerobatics (BA), night flight (NF) and basic instrument (BI) flight modules introduce new skill sets while providing an opportunity for pilot instructors to examine how cadets assimilate associated skills. Almost all of these modules are developed and consolidated during phase II (Cognac). A basic formation (BF) flying module is also introduced with the ultimate objective of allowing students to carry out and conduct sector missions and perform navigation tasks autonomously (i.e. solo flight without an instructor on board).

In the second half of 2018, the arrival of the Pilatus PC-21 during phase II (Cognac) training raised new challenges in terms of the technological changes brought about by an aircraft equipped with on-board simulation capability. Reassessing the structure of the training modules between phases I and II (or even III and IV), as well as a recap of the basic concept of flight, are necessary in order to ensure the perfect match between these new technological capacities and the skills of the flight crew.

## 8.4. Digitization of glass cockpits: what are the implications for pilot training?

### 8.4.1. *The basic technical skills revisited*

In the 2000s, general civil aviation saw the TAA revolutionize flight, training and safety customs. These aircraft, equipped with advanced technology, were expected to contribute to improving air safety. However, shortly after their introduction into the airspace, the U.S. authorities noted that the accident rate for this type of aircraft was higher than expected. To address this, the TAA-FITS research program developed by the FAA and a group of manufacturers was launched in 2003 [TAA 03]. One of the objectives of this program was to understand if and how the new technologies present in TAA were related to accidents involving this type of aircraft. The first elements resulting from this research show that modern TAA avionics requires an in-depth analysis of flying and learning to fly. The arrival of new technologies in these aircraft has thus generated significant changes in at least three areas:

– The first area concerns access to information and the management of information flows. Indeed, on board this type of aircraft and unlike what a pilot can find in conventional aircraft, there are multiple, continuous and abundant sources of information. Thus, the student pilot must develop skills at an early stage to be able to select the right information at the right time from all information flows, and to understand the scope and usefulness of this information in the context in order to make the right decisions.

– The second area affected by the change is well known to airline pilots and experienced military pilots. This is cockpit automation. Until now, a high level of automation (e.g. autopilot, FMS) has been reserved for commercial or military aircraft. Today, these systems are accessible and present in the TAA fleet of general aviation. However, many incidents and accidents in the recent history of aeronautics reflect the need for solid training to work in interaction with and within highly automated systems. It therefore seems essential that student pilots be made aware as soon as possible of the contributions, limits and dangers of modern cockpit automation in order to avoid the *ironies of automation* [BAI 83] observed in commercial aviation.

– Finally, the last area concerns the software and navigation architecture specific to computerized systems: "Systems Management" (SM). The management of a flight on board a TAA is mainly done via these systems. However, it is clear that their operating logic is more related to the world of information technology than to that of conventional aeronautics. Individuals unfamiliar with these "computer" logics may get lost in the many menus, submenus, options and features. This "computer" logic of the SM represents a danger because it has a strong propensity to capture the pilot's attention on the management of "system"-oriented epiphenomena at the expense of a more global flight management. The mastery of SM-related tasks becomes a key component of a young pilot's skill set.

These changes have a direct impact on how the pilot perceives and understands his environment, makes decisions and manages risk. The pilot's skills that are primarily affected by the observed changes are less about technical skills than about perception, sensemaking, decision-making and risk management.

#### 8.4.1.1. *Perception and sensemaking*

The prerequisite for any decision-making is to have a good representation of the situation. The construction of this "good" representation relies largely on the collection and synthesis of reliable, recent, accurate and multiple information (i.e. aircraft condition, location and nearby external environment). With the advent of modern avionics systems, all the information provided by the system to the pilot allows him to have a much more detailed and developed *situation awareness* (SA) (see Box 8.2 for a presentation of the concept). In response, this improvement in SA may cause the pilot to place a high degree of confidence in the system's ability to report the reality of the situation in a comprehensive and complete manner. The pilot could thus wrongly develop the feeling of "controlling" the uncertainty inherent in any flight through information control. However, it has been widely demonstrated that a high level of confidence in the systems can promote the natural tendency of pilots to "exit the loop" when placed in a passive situation of flight parameter supervision and/or autopilot supervision. Thus, self-checking procedures, for example, are reduced or even abandoned. These behaviors then give rise to "latent errors", which in the long term generally leads to a loss of situational awareness. This pitfall can be avoided if the student pilot is made aware at an early stage of defining when and why he or she should use a particular automatic system and how to stay in touch with the dynamics of a situation despite automation. Improving the quality of situational awareness is a step forward. Therefore, one would expect this to create value by systematically leading to an improvement in the quality of decision-making and risk management; however, this is not always the case.

In the case of the pilot, [END 96] defines situation awareness (SA) as "an internal model of the pilot with regards to the world around him" or more generally as "the perception of environmental events in time and space, the understanding of their meaning and the projection of their state into the near future". The author thus proposes a model for human decision-making expressed around three different types of SA: (i) the perception of events, (ii) the interpretation of the situation and (iii) the level of anticipation. In his model, Endsley also emphasizes that the quality of SA can be affected by the quantity of cognitive resources available. However, an individual's cognitive capacity is limited. The cognitive resources of the (young) pilot can be mobilized/saturated in different ways depending on the teaching method used. If these resources are saturated very early in the SA construction process (i.e. level 1), then the acquisition of skills related to situation interpretation (level 2) and anticipation (level 3 of Endsley's model), highly mobilized in the decision-making process, could be affected.

**Box 8.2.** *The concept of situational awareness*

#### 8.4.1.2. *Decision-making process*

Many accident reports[2] highlight that poor hazard assessment is often the cause of poor decision-making. Good decision-making, on the other hand, is often based on a good anticipation of events that could be problematic. As one pilot reminded us, "*a good decision... is above all about doing the right things at the right time*". Although modern aircraft systems provide the pilot with accurate and up-to-date information, the decision-making that results from this refined information does not mechanically reduce the uncertainty about the dynamics of the current situation [GOD 16]. It is not uncommon to observe the negative effects of the abundance and permanent updating of information on the pilot. The latter can develop a form of dependence on information that sometimes leads the pilot to focus on waiting for new information at the expense of taking action.

#### 8.4.1.3. *Risk management*

Information systems can sometimes have negative effects on pilot risk-taking; in some situations, for example, the abundance of information can lead to the pilot taking more risks than he would be willing to take in the

2 Reports available on the National Transport Safety Board (NTSB) website: https://www.ntsb.gov/_layouts/ntsb.aviation/index.aspx.

absence of this information. As one general civil aviation training pilot reported about weather-related risks, "*these information systems seem to create a feeling in the pilot that he is able to take a greater risk than his aircraft and experience allow*". Indeed, seeing the surrounding weather conditions on their MFD sometimes pushes pilots to go beyond the acceptable risk in relation to their level or experience. As an instructor points out: "*Just because you see in detail on your screen a weather danger zone, it does not give you the right to fly closer than the regulations allow*". It is obvious that the accuracy of the information made available to the pilot allows him, in specific and exceptional conditions, depending on the characteristics of the aircraft and his skills/experience, to fly closer to the "danger". However, this must remain an exception and not a rule. A few figures can attest to the reality of this problem. For example, 31% of TAA aircraft are involved in weather-related accidents compared to 4.7% for general aviation. The figures rise to 61.5% for fatal TAA aircraft accidents compared to 16.4% for the entire general aviation fleet. The implication of the above observation is that the management of a flight on board a TAA requires the student pilot to develop new skills early in their training. The introduction of the Cirrus SR20 in Salon-de-Provence (Ecole de l'air) also prompted the reassessment of flight training and education practices within the French Air Force. This has taken the form of a series of experiments to gather information on the effects of learning the "system basics" on the upholding of "flight basics" (see Box 8.3).

An experiment involving 10 pilot students from Ecole de l'air was conducted to verify, on the one hand, that they were able to quickly assimilate (i.e. one week) the operation and simple use of some advanced features of the SR20 system (i.e. some autopilot modes and GPS navigation on a map). On the other hand, this experiment also made it possible to verify that the in-flight use of these tools/features did not disrupt the normal operation of the aircraft or the fundamentals of flight, particularly following damage that rendered the use of the systems ineffective.

The experiment took place in two phases: the realization and evaluation of the effects of ground training on the one hand, and the results of flight tests on the other. The training was divided into two stages: an initial stage provided theoretical instruction mainly based on the system's technical manuals and on the Garmin PC trainer, allowing navigation through the system's various menus and submenus using a mouse and keyboard. The second training stage took place on an experimental flight simulator developed in-house (i.e. at the Air Force Research Center). The purpose of this step was to allow the student pilots to apply the theoretical knowledge acquired in a more realistic context by performing seven flight segments during which the use of the system's advanced functionalities varied. Each of these training steps ended with an assessment to verify the pilot's acquisition of knowledge and skills required to then participate in real flights. At the end of these two stages of training, the selected student pilots participated in two real flights spaced one week apart, during which they had full control and could use the systems studied on the ground. Instructor intervention was only required in the event of a flight safety hazard. The nature of these two flights (simple navigation flight) was identical to those performed using the simulator. These two flights made it possible to assess the ability of the student pilots to correctly mobilize the knowledge acquired in real conditions. They also tested their ability to return to flight fundamentals in the event of system failure (e.g. using the traditional "chrono-ground map" navigation method). For each flight, a debriefing form was completed by the instructors and students. The compilation of these forms made it possible to evaluate the students' flight performance, observe their ease of use of the systems management functionalities and gather comments from the crew.

Observation of the students' in-flight behavior and the use of debriefing sheets yielded interesting results. Concerning flight performance, no degradation was observed. The cadets seem perfectly capable of restoring in flight the SM techniques recently acquired on the ground while maintaining a high level of precision in maintaining standard flight parameters (i.e. heading, altitude, speed). The students also demonstrated a certain ease in the management of the systems studied by appropriately alternating between the different levels of automation and by detaching themselves from these pilot assistance systems when the situation required it. The return to conventional visual navigation, following system damage simulations, was generally well managed. The crew's (i.e. students + instructors) feeling towards early learning of the skills related to SM – the system basics – is rather positive. The only reservation was the instructors' concerns over extra time allocated to this new learning, which should not be at the expense of the other basics presented in section 3.2 (Table 8.1). These results are reassuring but not surprising since the student pilots who took part in this experiment were meticulously selected when they joined the French Air Force. It is not clear whether these results can be generalized to a more heterogeneous population, of the flying club type, facing the same technological changes.

**Box 8.3.** *Presentation of an experiment conducted at Ecole de l'air*

We have just outlined the challenges posed by modern avionics systems in flight training. The flight basics discussed (e.g. systems management, use of automation) represent the essential foundation of knowledge necessary to build the skills specific to a combat pilot. It is within this learning framework that embedded simulation can create more valuable learning situations, in particular thanks to the combination of two elements: first, the possibility of injecting "virtual" information directly into modern avionics systems, and second, the construction of a balanced training program, promoting the individual's progression during the different training phases.

### 8.4.2. *Reconfiguring the training toolset? The role of embedded simulation*

The term "Embedded Training" or "Embedded Simulation" is generally used to describe any training capability integrated or added to a system, subsystem or operational equipment [KEU 10]. A 2009 CHEAr report[3] stated that "the *concept of embedded simulation is to embed simulation functions in real systems*". This capability thus allows an operator to use his own operating system under the conditions for which it was designed, without the actual situation being available (i.e. the presence of an air/air or ground/air threat), or the use of the actual system being authorized (e.g. laser illumination, electromagnetic interference). In practice, embedded simulations materialize mainly through the generation of a virtual tactical environment in which fictitious threats, or events of all types, are simulated and presented on the interfaces of the real operating system. The use of this new educational tool obviously has effects on the quality of learning and air safety. In particular, the implications of on-board simulation on the processes of acquiring the combat pilot skills are twofold. On the one hand, it is a question of whether the level of performance achieved at the end of training will be higher (performance defined in terms of quality and/or speed of learning) than that obtained by current "traditional" teaching methods. On the other hand, it is a question of whether on-board simulation will better

---

3 Centre des hautes études de l'armement. Report of the 2nd Committee of the 45th National Session. Title: "Future of simulation for force training: what are the benefits for operation and what are the limits?"

prepare student pilots to deal with the complexity of the operational contract on modern weapon aircraft.

Current educational practices, taught on older generation aircraft (e.g. Alphajet), allow the acquisition of skills useful to combat pilots operating within a simple "imaginary" tactical framework. We use the term "imaginary" because everything happens "in the heads" of the crew. A training or instruction mission dedicated to learning evolution in a simple tactical environment begins with a ground preparation phase of the mission, by the instructor, on the one hand, and the student pilot, on the other hand. The former designs a credible tactical situation to develop tactical skills and learn to manage the dynamics of a situation. For example, for a mission to attack a target on the ground in a simple tactical setting, the instructor designs a situation that usually includes a target, a friendly/enemy line with associated schedules (i.e. time on target; altitude and gate time), as well as a set of ground-to-air and/or air-to-air threats, waiting and refueling zones and weather uncertainty zones. The student pilot must prepare his flight, taking into account all this information. He must define the main "route" on the objective, the alternative "routes" in case of diversion, taking into account the known threat (i.e. provided on the ground by the instructor). Once in flight, the student and the instructor "carry out" the planned mission. The instructor plays the role of an air traffic controller and can thus modify or adapt, by radio broadcasting, certain events or threats from the initial situation (e.g. appearance of a new threat, traffic in the area, deterioration of weather conditions). The student must process the new information provided by his instructor to adapt to the constraints and continue the mission. To achieve this, the student notes the important elements on his or her map (paper/pencil), as well as stores all or part of this information in his or her working memory in order to "make all the elements of the situation live in his or her head". Indeed, if he cannot visualize it on his on-board instrumentation, he must mentally calculate the dynamics of each entity to constantly update their spatial and temporal evolution. However, this mental simulation consumes cognitive resources. In a traditional learning context, it is not uncommon to see student pilots being overwhelmed very quickly by the evolution of the tactical situation in which they are engaged (e.g. the appearance of a new ground-to-air or air-to-air threat). This overwhelming sensation is very common among novices in all fields of application. An individual, unfamiliar with a situation in a particular field, or with the equipment and systems used, will be considerably overwhelmed by the collection and understanding of information, and the formulation of correct answers (i.e. the three levels of situational awareness, SA). More specifically, the research and collection of relevant information by a novice is likely to be suboptimal as it is highly demanding on cognitive resources. Some key information might sometimes be overlooked. For the student pilot, however, this phenomenon has the advantage of making him more aware of his cognitive limitations and the need to develop strategies to better manage the situation. From a pedagogical point of view, this practice also forces the young pilot to constantly adapt to the gradual increase in the complexity of training situations. This necessary adaptation will condition his entire training course.

While the pedagogical aim of current practices is not called into question, the method can reduce the student's progress rate regarding acquiring the combat pilot's skills. In particular, it is the development of SA adapted to the management of increasingly complex situations that could be lacking. With a traditional method, the dissemination of new tactical data is mainly oral. This requires the student to mentally calculate the dynamics of each element of the situation in order to have useful and usable information (i.e. the updated spatial and temporal position of the elements of the situation). With a classical method, the student is obliged to mentally "make" the information he needs to use so as to build a mental representation of the world around him. By immediately soliciting the student pilot's resources to "only" calculate and generate information useful for SA development, the risk is to limit the duration and depth of learning. On the one hand, this reduces the possibilities of collecting and managing a larger number of elements in a situation (SA level 1) because of the rapid saturation of the individual's resources induced by the mental simulation of each element. The level of complexity of the tactical situations proposed for the acquisition of combat pilot skills will therefore be limited. On the other hand, it reduces the amount of resources available to develop the skills required at levels 2 and 3 of situational awareness (i.e. interpretation and anticipation). However, studies conducted in the military field (i.e. infantry; [STR 01]) show that it is precisely these skills that are most lacking among young officers. It appears, for example, that in learning situations, which are characteristic of combat situations, junior officers have great difficulty (i) in merging information to obtain an overview of a situation (i.e. SA level 2) and (ii) in defining different action options (i.e. SA level 3). These steps (SA levels 2 and 3) require complex cognitive processes and knowledge. For novices in the field, correctly understanding the meaning of everything they perceive is difficult because of their limited experience in interpreting and anticipating this information. The authors of the study point out, for example, that junior officers have difficulty prioritizing tasks, understanding the impact of certain factors on troop fatigue or visualizing the appropriate positioning of soldiers to avoid fratricidal fire. Among the other difficulties identified, the authors also cite understanding the enemy's strengths and weaknesses, such as the enemy's strategic combat zones. With regard to level 3 of the SA, the young officers also have difficulty in (i) developing different alternative action plans, (ii) predicting the rapid use of ammunition and supplies and (iii) anticipating the enemy's intentions, or even in anticipating the consequences of the maneuvers of their own troops. The authors conclude their study by mentioning that the nature of the difficulties encountered by young infantry officers is in line with the problems of SA development also identified in other areas. Thus, they strongly recommend that the development and acquisition of levels 2 and 3 skills be major issues in future combat training programs.

With respect to current combat pilot training, the equipment (i.e. Alphajet training aircraft) and the conventional method may not be adequate to allow the development of SA levels 2 and 3. With current teaching methods (based on Grob or Alphajet), the student spends a lot of time mentally calculating the dynamics of a small number of events. These calculations consume cognitive resources that can no longer be mobilized to support higher-level cognitive processes involved in decision-making (SA levels 2 and 3). In the absence of modern on-board instrumentation capable of displaying a tactical situation (e.g. combat aircraft equipped with the Liaison 16 [L16; GOD 13]), tactical data exist only in the student's (and instructor's) head and only "live" at the cost of significant cognitive effort. Under these conditions, it is not possible, in the current training, to increase the complexity of missions without unnecessarily risking overloading the student and compromising air safety. However, modern avionics systems in general, and Embedded Training (ET) in particular, would enable the development of the initial phases of school training and optimize the acquisition of combat pilot skills. The on-board simulation capability, supported by modern avionics systems, allows a complete visual display of a fictitious tactical situation by allowing virtual and real elements to be mixed on the same interface. This hybrid display capability (real + virtual) gives the student pilot the opportunity to have visual access to an enriched, complete and direct synthetic representation of his environment. By physically materializing interface information, the student pilot is relieved of the need to mentally perform all the costly work of "making" and storing information useful for understanding the situation in which he or she is operating. The real-time display of information also provides the student pilot with direct access to level 1 of the SA (i.e. perception of the elements of a situation) with a sufficient level of accuracy and reliability, without significant resource costs. This is exactly what tactical data links such as Liaison 16 seem to offer in the latest generation of fighter aircraft.

Embedded simulation therefore provides a learning framework that shares several attributes of a real combat situation and can, under certain conditions, promote the acquisition of tactical, situational and informational skills. By allowing a reduction in the cognitive resources mobilized during level 1 SA, embedded simulation facilitates the development of the higher-level cognitive processes involved in interpreting and understanding increasingly complex tactical frameworks. To make sense of a situation in detail, it is indeed necessary to have a solid experiential basis on which to build. Studies carried out, in particular within the framework of the FAA-Industry Training Standards (FITS) TAA program, show that it is advisable to place the student pilot, very early and regularly during his training, in credible and realistic situations with regard to what the reality of a real flight will be. This is what is proposed by a learning method based on the "scripting" of the learning situation. By allowing the student pilot to assess the relevance of his situation analysis and to assess the consequences of his decisions and actions within these fictitious but realistic tactical frameworks, the probability of obtaining a positive learning transfer between combat training and reality is increased. Because it links the constraints of the simulated situation with those of a real situation, on-board simulation is likely to anchor the skills being acquired more deeply in memory. In the same manner, on-board simulation offers a learning environment conducive to the accumulation of experience, combining the advantages of classic real training (flight stress, physical sensations, acquisition of the sense of the air) with those of ground simulator training (i.e. safe and controlled learning environment, wealth of possible scenarios, identical repetition of a learning situation, level of complexity of the tactical situation, fewer aircraft "flying" into the air). Thus, from a learning point of view, it can be assumed that by materializing useful information (threats, routes, objectives, etc.) on the interfaces, embedded simulation gives the pilot access to a "virtual" reality capable of soliciting cognitive processes identical to those involved in a real exercise. Contrary to what is proposed in classical methods, the tactical situation here is no longer only imaginary: it becomes real through its (virtual) physical representation on the interfaces. Each decision taken, each action carried out, and each experience or knowledge acquired by the student pilot in on-board simulation will contribute to the development of an experiential base adapted to the conditions that will actually be encountered in operational missions. Thanks to on-board simulation, the student pilot will be able to mobilize more cognitive resources to quickly develop a situational awareness that will allow him to achieve a high level of performance in his combat pilot role. From a pedagogical point of view and in terms of the training curriculum, it becomes possible to expose student pilots to complex tactical frameworks at an earlier stage and, by extension, to support a high level of flight safety. By allowing the student pilot to experiment with learning conditions close to those of the real operational world, embedded simulation contributes to the development of efficient and safe behaviors.

## 8.5. Discussion and conclusion

The enabling conditions for an efficient and safe operation of the new capabilities offered by the latest generation aircraft are based on a reorganization of the initial phases of flight training (i.e. phases I and II) in order to harness the full benefits of their technological potential (e.g. systems management and the role of automation). The skills associated with such a reorganization are dual. On the one hand, they correspond to the technical fluency in the handling of modern avionics systems, and on the other hand, they refer to specific competencies related to military flights. Training must be thoroughly rethought by somewhat abandoning the traditional stacking pattern of abilities in which students first learn the basics of piloting, before exploring the "tactical" informational and situational capabilities of systems. In other words, the systems become an integral part of the use of modern aircraft and no longer differ from "normal" use. Training must therefore integrate these new capabilities on an ongoing basis and not only after the *ad hoc* acquisition of certain flight capabilities.

This contribution shows how much the concept of "flight basics" has changed. In a traditional training course, the focus is too often positioned only on the acquisition of "flight basics" insofar as the essential skills to fly the aircraft are mainly in the field of understanding the "physical" behavior of the aircraft (no modern systems on-board). Pilots trained on conventional aircraft in school and then on older generation weapon aircraft are well aware of the importance of this area of expertise. A former fighter pilot patented on Mirage F1 told us that "*at the time, you had to have very good flying skills to fly this type of aircraft, you had to be ultra-efficient if you didn't want to have an accident, because there weren't as many systems and piloting aids as there are today*". However, the evolution of on-board technology, decision support systems and cockpit automation have contributed to the need to enlarge the definition of "basics". There is a growing awareness on the part of trainers that the acquisition of certain skills, which are taught too late, can be achieved earlier in the training curriculum.

Finally, the concept of "basics" must be understood in two dimensions. The first concerns the fundamental skills that allow the pilot to perform a simple solo flight on a modern aircraft. While the objectives have not changed (e.g. safety, knowledge of the aircraft's physical behavior, navigation, etc.), the scope of the initial basic skills is now more extensive on modern aircraft than it was on conventional aircraft. In the final report of the TAA Safety Study Team, the authors distinguish three categories of basic skills: "Physical Airplane", "Mental Airplane" and "Risk Assessment and Management". While the first category is based on traditional basic skills ("stick-and-rudder"), the latter two are directly influenced by the capabilities offered by new aircraft equipped with glass cockpits (see Box 8.4 for an overview).

"Physical Airplane" refers to the frame and flight controls (i.e. flaps, landing gear, lights, etc.) used to launch or land an aircraft whose performance is characteristic of conventional high-wing aircraft in civil flying clubs. The skills involved are the traditional basic stick-and-rudder skills. The authors of the study believe that a failure in acquiring these skills leads to fatal accidents, relatively rarely today, but can be very serious accidents or incidents. In addition, this type of failure can also interfere with the pilot's ability to control the "Mental Airplane". The "Mental Airplane" concerns the combination of the management of modern avionics systems – used for communication, navigation and surveillance – and flight management. The "Mental Airplane" includes, on the one hand, the ability to use the key functions of each technical subsystem individually. It also includes how to use these functions correctly and at the right time in the current operation. The basic skills involved in the "Mental Airplane" field then concern both the pilot's ability to understand the usefulness of modern avionics systems in improving flight safety and his ability to understand the limitations inherent in these modern systems. Indeed, an overestimation of the capacities of the systems or a misuse of these systems can lead to a decline in the level of safety. Finally, the basic skills must also allow the student pilot to understand that safety is the result of a precarious balance between the capabilities and limitations of the pilot/aircraft pair and the requirements of the mission. The last critical area identified by the authors of the report concerns "Risk Assessment" and "Risk Management". The acquisition of skills related to this field is largely dependent on the good mastery of the skills acquired in the other two areas previously described. However, it is also through the development of decision-making skills in situations that risk assessment and management skills improve. A training program based on scenarios written and constructed with specific and targeted pedagogical objectives ("Scenario-based Training") is strongly recommended to prepare pilots for the specificity of managing a flight on TAA [SUM 07]. Pilots are thus regularly confronted with realistic "controlled" critical situations, enabling them to acquire the experience required for decision-making in a dynamic situation on modern aircraft. Another advantage of a scenario-based training program is to teach/demonstrate to the pilot that increasing the capabilities of modern aircraft does not automatically result in an improvement in the pilot's ability to safely perform all missions.

**Box 8.4.** *The typology of the basics according to the TAA Safety Study Team*

The second dimension concerns the high level of performance required when acquiring the different types of initial basic skills because, in our opinion, they condition all the learning achieved in the following phases.

For example, the mastery of SM and the use of automation are essential prerequisites for acquiring the specific skills of the combat pilot operating in a complex tactical environment. Although the new aircraft are now deployed in the units, there is still a strong belief among the pilot community and training managers that training conducted too early on aircraft equipped with modern avionics could affect the acquisition of flight fundamentals. In reality, it is not so much the fundamentals in the strict sense that change, but rather the interface and systems (systems management) through which the pilot takes information and acts that differ and modify certain practices. Despite significant differences in the way a modern training aircraft (e.g. TAA) is piloted/managed, most of the risk factors inherent in flight are the same on "digital" and "analog" aircraft. Consequently, becoming aware of the risks and opportunities offered by these new interfaces and systems too late may constrain the acquisition of flight fundamentals.

In order to avoid this pitfall, the syllabuses of the first training phases must take into account the specificity of the new modes of interaction with modern avionics to enable pilot students to develop a new set of skills (i.e. "systems management basics") in the field of information management and systems management at a very early stage. The on-board simulation capability is an excellent means of developing the combat pilot's skills [ROE 09], in particular by bringing the learning environment closer to the real work environment ("Train as you fight"). Again, the efficient and safe use of this simulation capability depends on the level of preparation of the student pilots in the early phases of training. The arrival of this new capacity must necessarily lead to a profound structural reform of the learning content during the various phases of the pilot's progression. A new distribution of learning objectives at the different stages of the training process therefore appears essential.

## 8.6. References

[ADL 91] ADLER P., "Workers and flexible manufacturing systems: Three installations compared", *Journal of Organizational Behavior*, vol. 12, pp. 447–460, 1991.

[BAR 08] BARBAROUX P., GODÉ C., MERINDOL V., VERSAILLES D., Les Compétences de l'Armée de l'air à horizon 2015, État-major de l'Armée de l'air, Bureau Plan, p. 125, 2008.

[BAR 10a] BARBAROUX P., GODÉ C., Maintien et Transfert des compétences dans les escadrons de transport tactiques: du Transall à l'A400M, État-major de l'Armée de l'air, Bureau Plan, p. 105, 2010.

[BAR 10b] BARBAROUX P., GODÉ C., "Quelle typologie pour identifier les compétences ? Le cas des pilotes de transport de l'armée de l'air", *Information Science and Decision Making*, no. 40, 2010. Available at: http://isdm.univ-tln.fr/.

[BAR 11] BARBAROUX P., "Technologie polyfonctionnelle et compétences des acteurs : le cas des pilotes de chasse de l'armée de l'air", *Revue Française de Gestion*, vol. 37, no. 212, pp. 29–43, 2011.

[BAR 12] BARBAROUX P., GODÉ C., "Changement technologique et transfert de compétences : une réflexion à partir du cas des équipages de transport de l'armée de l'air", *Management International*, vol. 16, no. spécial, pp. 57–73, 2012.

[BAR 16] BARBAROUX P., GODÉ C., "Le briefing-débriefing: une procédure pour lever les barrières pesant sur l'apprentissage organisationnel ?", Gérer & Comprendre, 2016/2, no. 124, pp. 41–51, 2016.

[BAI 83] BAINBRIDGE L., "Ironies of automation Automatica", vol. 19, pp. 775–779, 1983.

[BRE 02] BRESNAHAN T.F., E. BRYNJOLFSSON L., HITT M., "Information technology, workplace organization, and the demand for skilled labor: Firm-level evidence", *The Quarterly Journal of Economics*, vol. 117, no. 1, pp. 339–376, 2002.

[BOL 09] BOLAND R., THIEL D., "A special issue on simulation", *Systèmes d'Information & Management*, vol. 14, no. 4, pp. 5–7, 2009.

[CAM 12] CAMACHON C., BLÄTTLER C., La simulation sol et la simulation embarquée : Enjeux pour la formation et l'entrainement du personnel navigant, État-major de l'Armée de l'air, Bureau Plan, p. 60, 2012.

[CAM 14] CAMACHON C., BLÄTTLER C., Impact de l'introduction de modules de formation à la Gestion de Systèmes (GS) sur l'apprentissage des basiques de vols en Phase I. Contrat de recherche sur commande des Ecoles de Formation du Personnel Navigant (E.F.P.N), Note de synthèse, p. 6, 2014.

[CAP 93] CAPPELLI P., "Are skill requirements rising? Evidence from production and clerical jobs", *Industrial and Labor Relations Review*, vol. 46, no. 3, pp. 515–530, 1993.

[DUB 15] DUBOIS E., BLÄTTLER C., CAMACHON C., HURTER C., "Eye movements data processing for Ab initio military pilot training", in *Intelligent Decision Technologies*, pp. 125–135, Springer International Publishing, London, 2015.

[END 96] ENDSLEY M.R., "Automation and situation awareness", in *Human Factors in Transportation. Automation and Human Performance: Theory and Applications*, PARASURAMAN R., MOULOUA M. (eds), pp. 163–181, Lawrence Erlbaum Associates, Inc., Hillsdale, 1996.

[FUN 99] FUNK K., LYALL B., WILSON J., VINT R., NIEMCZYK M., SUROTEGUH C., OWEN G., "Flight deck automation issues", *The International Journal of Aviation Psychology*, vol. 9, no. 2, pp. 109–123, 1999.

[GAL 02] GALE H.F., WOJAN T.R., OLMSTED J.C., "Skills, manufacturing technology, and work organization", *Industrial Relations*, vol. 41, no. 1, pp. 48–79, 2002.

[GOD 10] GODÉ C., BARBAROUX P., "La fabrique des usages technologiques en environnement volatil", *Management & Avenir*, no. 32, pp. 71–90, 2010.

[GOD 13] GODÉ C., LEBRATY J.F., "Improving decision making in extreme environment: The case of a military Decision Support System", *International Journal of Technology and Human Interaction*, vol. 9, no. 2, 2013.

[GOD 16] GODÉ C., BARBAROUX P., "Combining technologies' properties to cope with uncertainty: Lessons from the military", *International Journal of E-Entrepreneurship and Innovation*, vol. 6, no. 1, pp. 1–18.

[GOL 98] GOLDIN C., KATZ L.F., "The origins of technology-skill complementarity", *The Quarterly Journal of Economics*, vol. 113, no. 3, pp. 693–732, 1998.

[KEU 10] KEUNING M.F.R., "Embedded Training and LVC", *Proceedings of the NATO MSG-076*, 2010.

[LEI 96] LEI D., HITT M.A., GOLDHAR J.D., "Advanced manufacturing technology: Organizational design and strategic flexibility", *Organization Studies*, vol. 17, no. 3, pp. 501–523, 1996.

[NOK 09] NOKES T., "Mechanisms of knowledge transfer", *Thinking & Reasoning*, vol. 15, no. 1, pp. 1–36, 2009.

[OIR 05] OIRY E., "Qualification et compétence : deux sœurs jumelles ?", *Revue Française de Gestion*, no. 158, pp. 13–34, 2005.

[ROE 09] ROESSINGH J.J.M., VERHAAF G.G., "Training effectiveness of embedded training in a (multi) fighter environment", in *Proceedings of the NATO HFM-169 Research Workshop on the Human Dimensions in Embedded Virtual Simulation. 3:1-3: 8,* Orlando, FL., 20–22 October, 2009.

[SAL 98] SALAS E., BOWERS C.A., PRINCE C., "Special issue: Simulation and training in aviation", *International Journal of Aviation Psychology*, vol. 8, no. 3, 1998.

[SAL 09] SALAS E., WILDMAN J.L., PICCOLO R.F., "Using simulation-based training to enhance management education", *Academy of Management Learning & Education*, vol. 8, no. 4, pp. 559–573, 2009.

[STR 01] STRATER L., JONES D.G., ENDSLEY M.R., "Analysis of infantry situation awareness training requirements", Marietta, GA., SA Technologies, 2001.

[SUM 07] SUMMERS M.M., Scenari-based Training in TAA as a method to improve risk management, Technical Report – Embry-Riddle Aeronautical University, 2007. Available at: http://citeseerx.ist.psu.edu/viewdoc/download?doi=10.1.1.573.4527&rep=rep1&type=pdf.

[FAA 03] FAA – Industry training standards (FITS) program plan, Government Printing Office, Washington, 2003.

[YIN 03] YIN R., *Case Study Research: Design and Methods*, Sage Publication, Third Edition, 2003.

# 9

# Towards the Advent of High-Altitude Pseudo-Satellites (HAPS)

ABSTRACT. Initiated in the 1950s with the invention of photovoltaic cells, almost unlimited endurance high-altitude solar drone concept is still under heavy development because of persistent technological barriers. The low propelling power extracted from solar source involves a specific design of the airframe consisting of a very lightweight and flexible airframe particularly vulnerable to destructive interactions. The latter fall within the scope of aeroelasticity, a discipline between aerodynamic and structural mechanics. This chapter illustrates the transdisciplinary approach required to achieve such a challenging goal and presents innovative solutions based on research work. These should enable commercial exploitation of such a concept and then fill the gap between conventional HALE drone and satellites in the field of observation and telecommunication.

## 9.1. Introduction

The history of aviation has been defined by successive advances that have made it possible, step by step, to improve the performance of heavier-than-air aircraft. Whether in terms of range, flight speed or cruising altitude, technological advances and pioneers ready to implement them have pushed the boundaries of aviation to new limits over the years. Among the limitations, endurance poses a number of challenges for researchers and aircraft designers alike.

The endurance of an aircraft is in itself a fairly simple concept: it is the time that elapses between the time the wheels (when there are wheels) leave the ground and the time when contact with the ground is re-established.

Chapter written by Bertrand KIRSCH and Olivier MONTAGNIER.

This endurance is directly related to the amount and type of energy available for the engine. Fossil fuels, such as kerosene, are almost the only ones used in aeronautics because of their high energy densities (Figure 9.1). The major disadvantage of this solution is that it does not make it possible to regenerate, even partially, the quantity of energy on board at the start, except by using in-flight refueling. This technique, which is fundamental in military aeronautics, however, has at least two major limitations in terms of optimizing endurance. On the one hand, it does not increase the intrinsic endurance of an aircraft and thus transfers the "problem" to tanker aircraft. On the other hand, it forces refueling and refueled aircraft to be able to fly in the same conditions, particularly in terms of speed and altitude.

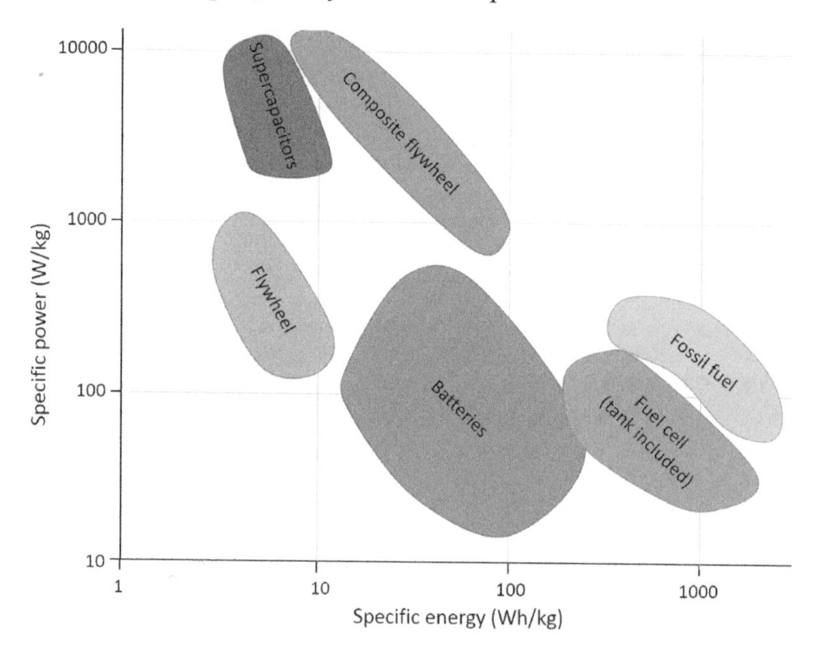


**Figure 9.1.** *Comparison of the energy sources available for drones (Ragone diagram): specific power versus specific mass in logarithmic scale. For a color version of the figures in this chapter see www.iste.co.uk/barbaroux/technology.zip*

Electricity or hydrogen can be an alternative to kerosene. In the case of aircraft, electricity is stored in batteries and directly used by one or more electric motors running one or more propellers (e.g. the Airbus E-Fan that flew in 2014 [JUV 16]). Hydrogen is used either as a fuel in a hydrogen engine (e.g. Tupolev Tu-155 stolen in 1989) or as an energy source for a fuel cell generating electricity to replace the battery in the electrical solution (e.g. ENFICA-FC developed by Politecnico di Torino [ROM 13]). This hydrogen is stored in tanks under very high pressure to minimize its volume. Unfortunately, these different alternatives, however interesting they may be from an ecological point of view, will not solve the endurance problem because they have a lower energy density than kerosene (Figure 9.1).

Solar energy seems to be the only solution to solve the limitation of endurance. This energy is unlimited during the day for an aircraft flying above the cloud cover, i.e. above the troposphere (between 0 and 13 km). The operating principle is then quite simple. During the day, photovoltaic cells positioned on the surface of the aircraft recover the energy required for flight and instruments, to which the energy required for night flight must be added. The latter is stored either in batteries or as hydrogen, using a reversible (also called regenerative) fuel cell. At night, the stored energy is then returned for use by propulsion and on-board instruments. Such a concept, when implemented in a UAV system to avoid flight time limitations, improves the endurance limit no longer in terms of on-board energy autonomy but to the time allocated between two maintenance visits, which can thus be up to several years. This advanced optimization of the HALE (High-Altitude Long-Endurance) solar UAV is a technological implementation of the more general concept of High-Altitude Pseudo-Satellite (HAPS). These solutions have been studied since the mid-1980s. However, to date, no such devices are in operation.

The objective of this chapter is to analyze the design challenges of these solar HALE UAVs to explain the time elapsed between the creation of the concept, its implementation and perhaps one day its exploitation and thus the completion of the innovation process. To do this, we will first present the challenges with regard to applications and capacity and then the history of solar aeronautic projects, records and accidents. We will then look at the persistent technological barriers that have so far prevented the full implementation of this aircraft concept. In particular, we will show the need to design flexible composite wings and the physical consequences that result from this need, such as dynamic instability (also known as flutter). Finally, among the possible avenues for development towards a viable solution, we will mention an idea directly resulting from research: aeroelastic tailoring.

## 9.2. Capability issues: observation and telecommunications

The potential of HAPS is great because it is mainly a question of extending the use of the aircraft in a capacity domain previously reserved for satellites. Here, it is not a question of globally replacing one means with another but rather filling a capacity gap, hence the term pseudo-satellites. Where the installation of heavy space asset will be expensive and unsuitable, the implementation of a high-altitude drone, which could take a few dozen hours to reach the area, would make sense, provided that it can then remain in the area. Two main applications that concern both civilian and military users stand out: observation and telecommunications.

Indeed, by flying at high altitude and with extremely high endurance, these aircraft have no equivalents. The targeted altitudes (about 20 km) are close to the maximum cruising altitude of the Lockheed U-2 reconnaissance aircraft and are ideal for observation purposes (Figure 9.2). Here, the drone will not be very vulnerable as it is located in an area totally free of air traffic, with airplanes flying at an altitude of approximately 12 km. Technologically, the advantage of these altitudes is being above the clouds, reducing the atmospheric attenuation of solar energy and being located in a zone with light winds (facilitating the durability in this zone). On the other hand, the very low air density is very unfavorable in terms of lifting the device.

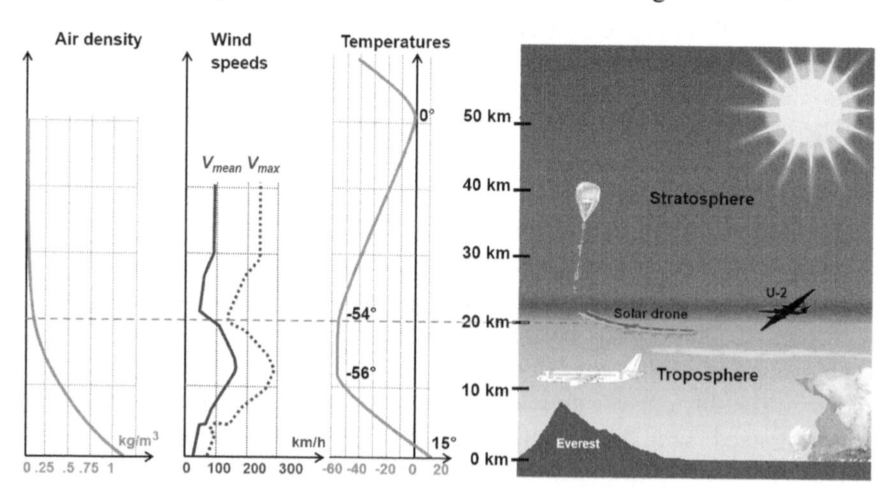


**Figure 9.2.** *Positioning of a HAPS in the atmosphere*

At an altitude of 20 km, despite the roundness of the Earth, a HAPS could function over a radius of 200 km. With a suitable transmitter, as few as four to five platforms could cover the entire French territory (Figure 9.3). According to Cestino [CES 06], a HAPS of this type would replace five conventional energy MALE (Medium-Altitude Long-Endurance) UAVs at a cost four times lower (750 euros per flight hour vs. 3060 euros per flight hour).

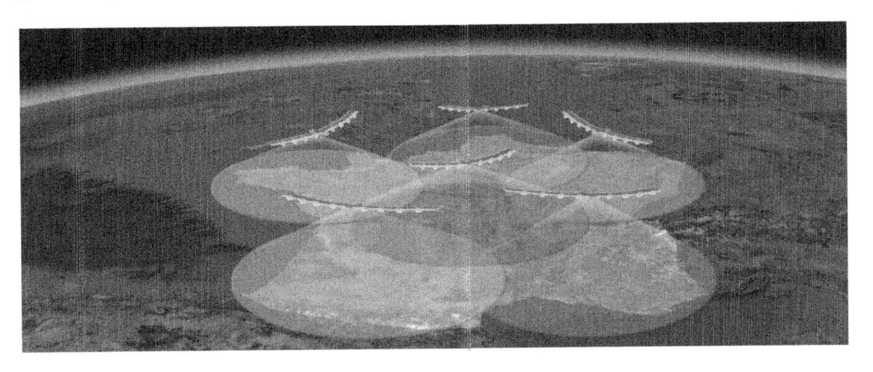

**Figure 9.3.** *Five solar HALE UAVs positioned at an altitude of 20 km could cover the whole of France*

NASA has identified potential applications in various fields. In the earth sciences, it is estimated that this type of drone could be used for meteorology, oceanic and geological observations, and monitoring of air pollutants and the ozone layer. As far as defence and security applications are concerned, these altitudes would make it possible to set up telecommunications relays applied to the surveillance of borders, maritime areas, the illegal transport of radioactive materials, the monitoring of forest fires, etc. Finally, in terms of commercial applications other than telecommunications relays, these drones could be used for monitoring road transport, natural resources, topographical surveys and precision agriculture.

## 9.3. Solar flight history: projects, records and accidents

Photovoltaic cells were discovered in 1954. The same year, August Raspet proposed to install these cells on the top surface of a glider and to use an electric propulsion system. The idea of solar flight was born, marking the beginning of a long phase of technological development.

In 1974, the Sunrise I UAV, with a wingspan of 9.7 m and a mass of 12 kg, made its first solar flight in California. In 1975, its successor, the Sunrise II, reached a record altitude of 5.2 km [BOU 85]. The development of the first solar drones was then closely linked to the development of solar aircraft such as Solar One (GB), Solar Riser (US) and Solair I (DE). The Gossamer Penguin developed by AeroVironment (US) was the first aircraft to fly solely on solar energy (Figure 9.4). It made a 3 km flight to California on August 7, 1980. From this project, the Solar Challenger was born. It had a span of 14.3 m and an empty mass of only 90 kg [MAC 83]. Carrying a larger number of cells on its wings and rear, this uniquely solar aircraft was the first to travel long distances. It crossed the English Channel in 5 hours and 23 minutes on July 7, 1981. All these projects show how the mass is a significant factor in the amount of energy available. The wings are made of composite materials and covered with extra lightweight polymers (Figure 9.4).

**Figure 9.4.** *Gossamer Penguin during tests in 1980*

In the early 1990s, NASA launched the first major solar HALE UAV program called ERAST (Figure 9.5(a)). The objective was to build a drone capable of flying continuously for six months beyond 18 km altitude. The Pathfinder UAV made its first flights in 1994 [FLI 98]. It was actually built in 1983 by AeroVironment during a classified defence project. Pathfinder reached a record altitude of 15.4 km above California on September 11, 1995. From 1997, the tests were carried out in Hawaii, particularly due to its favorable sunshine and weather predictability conditions. It then obtained the record altitude for a propeller-powered aircraft (21.8 km). The efficiency of the cells then was only 14%, and the small batteries did not allow night flight. In 1998, Pathfinder became Pathfinder Plus (Figure 9.5(a) and 9.5(b)). The central part of the UAV wing was extended, and two engines were added. The potentially available power increased from 7500 W to 12,500 W. It set a new record altitude of 24.4 km on August 6, 1998; the flight lasted nearly 15 hours. In 2001, Pathfinder Plus was used to demonstrate its ability to carry out an irrigation control mission on the largest coffee plantation in the USA on the island of Kauai [HER 02]. Its payload was 45 kg. In 1998, the Centurion UAV (Figure 9.5(a)) was built and prefigured the Helios UAV, which was equipped with a reversible fuel cell for night flight. The latter made its first flights in 1999 (Figures 9.5(a) and 9.5(c)). It had a wingspan of 75 m, larger than a Boeing 747 with a total weight of less than 930 kg. Its wings were covered with 62,000 photovoltaic cells and its payload was approximately 100 kg. Its speed was only 35 km/h at sea level, but this reached well above 100 km/h at high altitude and even 270 km/h at very high altitude. Helios reached an altitude of 29.5 km on August 13, 2001. Unfortunately, during its second flight in a long-term configuration (i.e. with a fuel cell) on June 26, 2003, the drone encountered an unstable oscillation that proved fatal (Figure 9.6). The inquiry commission concluded that a pitch oscillation mode coupled with a structural mode [NOL 04] had appeared. In other words, the drone started flapping its wings when it was obviously not supposed to. This failure ended the ERAST program ($97 million mainly used for the development of solar drones) before the long-term flight could be evaluated.

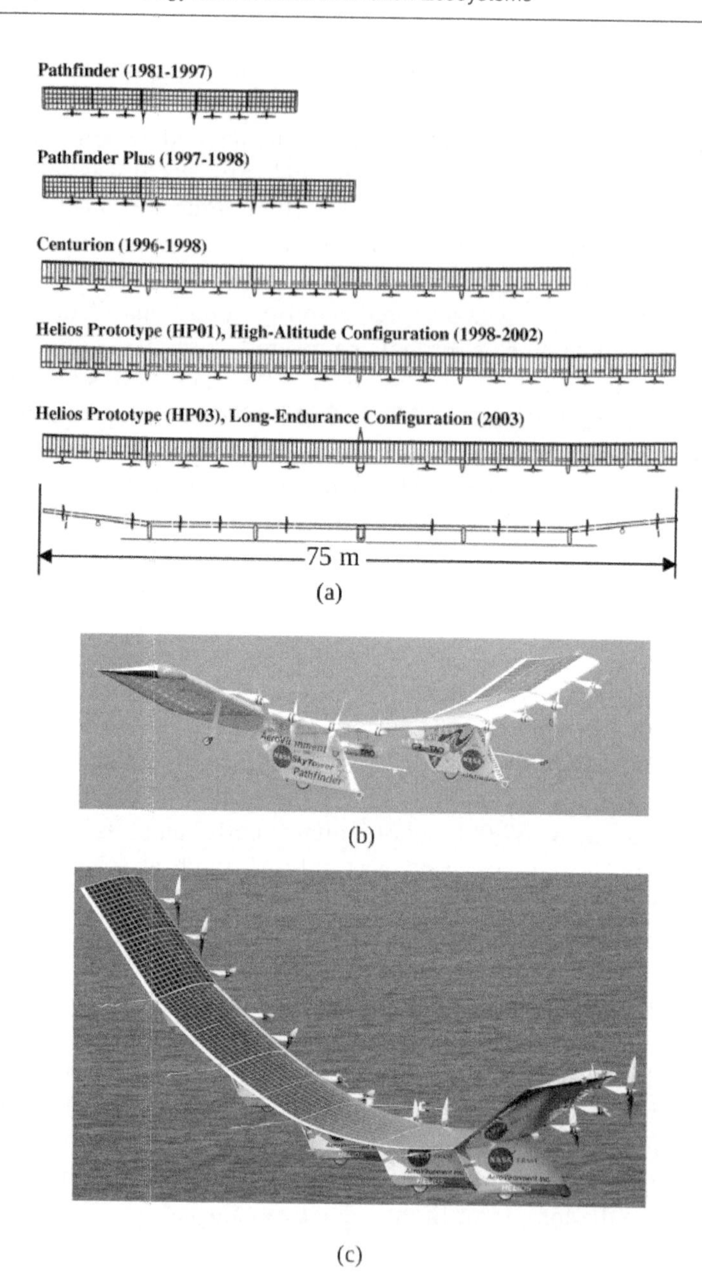


**Figure 9.5.** *NASA ERAST program: (a) comparison of the different solar drones, (b) Pathfinder plus in June 2002 and (c) Helios in July 2001*

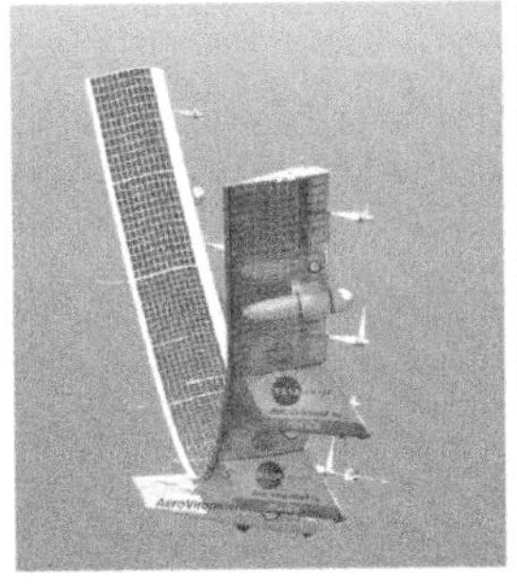
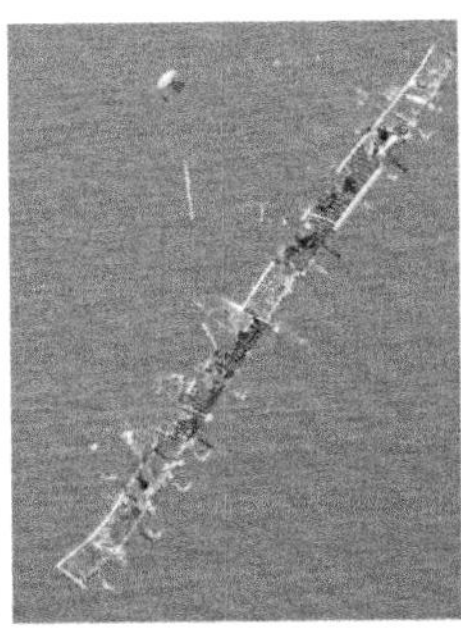
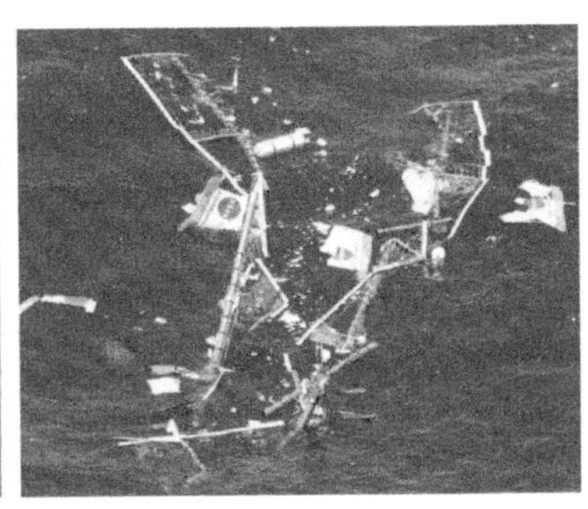

**Figure 9.6.** *Helios accident on June 26, 2003 (photos taken by the following helicopter)*

Over the last decade, several solar HALE UAVs and solar aircraft projects have been launched. In Europe, the most successful project is the Zephyr UAV (22.5 m wingspan for 53 kg for version 7) developed by Qinetiq and then acquired by Airbus Defence and Space (Zephyr S version also called Zephyr 7 shown in Figure 9.7). This UAV is relatively small and has a payload of only 2.5 kg. The latter is currently the holder of the endurance record with 14 consecutive days in flight (336 hours 22 minutes and 8 seconds from July 9 to 23, 2010). A new performance of 25 days 23 hours 57 minutes (and 22,589 m altitude) achieved in August 2018 by the Zephyr 8 model, whose characteristics have not yet been communicated, is currently being validated by the International Aeronautical Federation[1]. By means of comparison, the record for manned flight without refueling is held by the Rutan Voyager at 216 hours 6 minutes and 43 seconds from December 13 to 23, 1986, followed by the Solar Impulse 2, which achieved a performance without fossil fuels for 117 hours and 51 minutes from June 28 to July 3, 2015. It should be noted that Solar Impulse 2 is capable of continuous flights (days and nights) but with ideal weather (few clouds) because its cruising altitude is 8500 m.

In the USA, three major HALE solar UAV projects have been launched with varying degrees of success. The first being the Vulture projects launched by Darpa in 2008 for 90 million dollars. The objective was to transfer the knowledge acquired by the ERAST program to industrialists. The specifications were extremely ambitious since they were intended to produce a prototype capable of flying for five years without interruption. In 2010,

1 https://fai.org/records.

Boeing won the first phase for projects with its SolarEagle UAV against Lockheed Martin and Aurora Flight Sciences. Unfortunately, the project was abandoned in 2012. Another major project was Facebook's Aquila UAV (30 m span for 400 kg) developed by its Connectivity Lab and built by Ascenta in England (Figure 9.7). The company's objective was to be able to distribute Internet to remote areas (within an 80 km radius of the drone). The novelty of this aircraft was to be positioned at high altitude by helium balloons before starting to fly and then landing after three months by sliding on grass. The drone made its first flight on June 28, 2016 but crashed on landing, this time due to an interaction between the automatic flight controls and the flexibility of the wing [NTS 17a]. After the accident, the prototype was modified, and it subsequently completed a new 46-minute flight in 2017. However, on June 27, 2018, Facebook announced that it would stop developing the Aquila UAV to focus on the telecommunications aspects of this type of platform[2]. Finally, the third project was the Solara 50 drone (50 m long and 50 m wide) of the start-up Titan Aerospace, acquired by Google in April 2014 (Figure 9.7). The small-scale prototype crashed on its first flight on May 1, 2015, shortly after takeoff. The investigation report concluded that the speed was excessive, triggered by strong updrafts and a large deformation of the wing, leading to an in-flight failure of the left wing [NTS 10b].

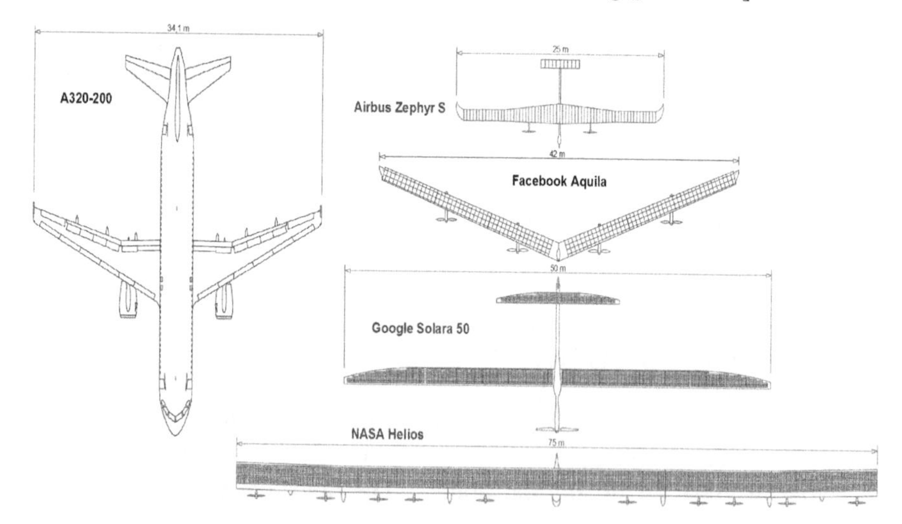


**Figure 9.7.** *Comparison of the different geometries of solar drones with an Airbus A320-200 (top view)*

2 https://code.fb.com/connectivity/high-altitude-connectivity-the-next-chapter/.

Finally, in China, the China Aerospace Science and Technology Corporation (CASC) developed a prototype of a solar HALE UAV called Rainbow (45 m span). On May 24, 2017, it flew for 15 hours and reached an altitude of 20,000 m.

A more complete history of HALE solar drones is given in [NOT 08] and [MON 13].

The concept of a heavier-than-air aircraft with unlimited endurance, due to the significant research efforts still required and the numerous accidents, is competing with the use of high-altitude airships. To illustrate this competition, let us mention Google, which abandoned the Solara 50 solar UAV project to focus on developing an aerostat-based solution. The company's efforts are now focused on the Loon stratospheric balloon project to provide network coverage in unserved areas. These balloons are 15 meters in diameter, fly at 20 kilometers above sea level and have a target endurance of 187 days. They were used in October 2017 to restore emergency network coverage to Puerto Rico residents after Hurricane Maria. The major disadvantage of this solution is that it is dependent on winds for direction, the only control parameter being the change of altitude.

Another lighter-than-air candidate for this endurance race is the Stratobus project of Thales Alenia Space, a 100-meter long airship designed to be placed at an altitude of 20 kilometers [BAU 15]. This platform has the advantage of being powered by an electric solution by means of a reversible fuel cell, which makes it repositionable unlike a stratospheric balloon. The announced endurance is 5 years; the first flight is expected to occur in 2019 with commissioning in 2020.

## 9.4. Resolution of a scientific and technological paradox

Reviewing the history of solar flight highlights the existence of a scientific and technological paradox. Indeed, despite the significant state and private investments already made in this field, the concept of a solar drone with unlimited endurance has still not been achieved, even though it implements technologies that have all made considerable progress individually over the past decades. In order to solve this paradox, the rest of this chapter focuses on the technological obstacles inherent in the concept of the solar drone.

### 9.4.1. *Solar energy: unlimited?*

The first question to ask when sizing a system that only runs on solar energy is whether it is available in sufficient quantities. This energy is unlimited during the day, and it is enough to use photovoltaic cells to recover it. Unfortunately, it is not that simple. The amount of energy that will be recovered will depend directly on the surface area of the cells and their efficiency, as well as on the position of the drone around the globe and the date it is located there. Indeed, the energy recovered depends directly on the angle that the light rays make with the cells (*a priori* horizontally). Figure 9.8 clearly shows that the power available in a day at our latitudes is highly dependent on the day of the year and is very low. For example, the power captured at noon at the winter solstice by a square meter of cells with an efficiency of 19% (corresponding to those arranged on Helios) is 80 W. In particular, the energy per square meter captured in a day, i.e. the area under the black curve, is five times lower in the winter solstice than in the summer solstice.

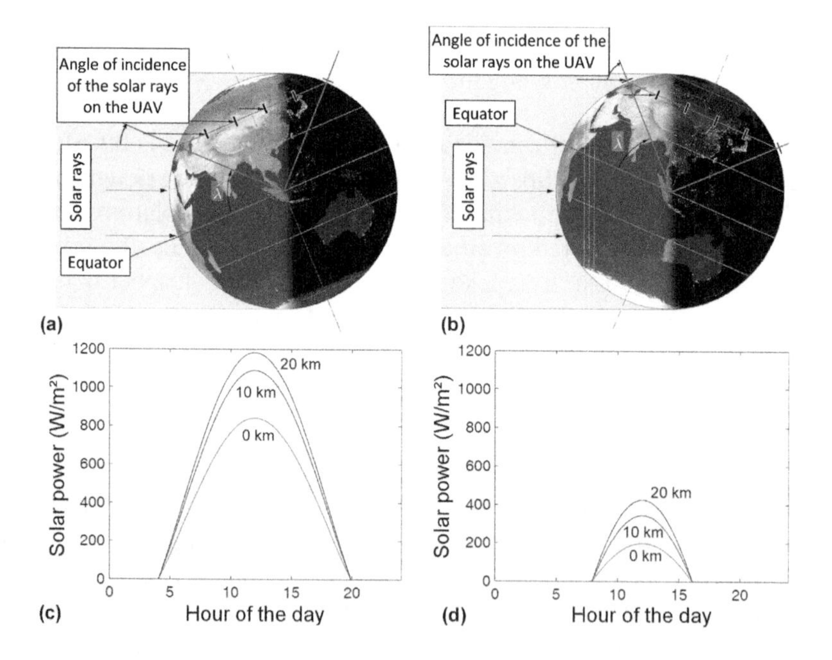


**Figure 9.8.** *Sunlighting of a solar HALE drone (a) at the summer solstice and (b) at the winter solstice and surface solar power available horizontally as a function of altitude, at 48° latitude (Paris), (c) at the summer solstice and (d) at the winter solstice*

Let us imagine now that we want to fly a conventional drone with solar energy alone. Let us take the example of a HALE observation drone such as the "Global Hawk" with a total mass of about 10 t and moving at 640 km/h at an altitude of 20 km. Maintaining this UAV in flight at its cruising altitude requires a propulsive power of about 610 kW. To obtain this power on the day of the summer solstice at noon and at our latitudes, assuming that 20% of the solar energy collected is converted into power, more than 2600 $m^2$ of solar cells (or the equivalent of half a football stadium) would be required. During the winter solstice and at noon, this area would reach nearly 7200 $m^2$. The latter would still be greatly increased if the drone were to continue to fly at night because a large part of the energy would have to be stored in batteries. In practice, the wing area of the "Global Hawk" is only about 50 $m^2$.

This demonstration quickly leads to the conclusion that a solar drone must maximize cell surface area, use high-efficiency cells but with a low surface mass and consume little energy.

### 9.4.2. *The keys to endurance*

Let us now return to the notion of endurance. From an aerodynamic point of view, the key parameter is the lift-to-drag ratio. As its name suggests, this is the ratio between the lift force and the drag force. The type of aircraft that optimizes this parameter is the competition glider, which can exceed 70 in ideal conditions. In practice, this means that in calm air, such a glider can move horizontally 70 km before landing from an altitude of one kilometer.

From a structural point of view, it is necessary to build a structure as light as possible for two main reasons. The first is to limit as much as possible the propulsive power used to compensate for losses due to aerodynamic drag. Indeed, the lighter the structure, the lower the balance speed to maintain altitude and the lower the drag. The second is to make as much mass available as possible to carry energy storage capacity and payload (sensors, relays, etc.).

The points mentioned above allow us to define a maximum endurance aircraft: a flying wing to maximize the surface area of solar cells, similar to a glider wing to optimize lift-to-drag ratio, flying at high altitude to maximize the energy captured and built using composite materials such as carbon fiber/epoxy resin to combine strength and lightness [MON 09; MON 10]. Almost all examples in part 3 follow this formula.

The definition of a maximum endurance aircraft and the associated technologies already seems to be defined in the aeronautics world; the use of composite materials in aircraft manufacturing, for example, is becoming increasingly important in military and civil aeronautics. It exceeds 50% by weight on the Airbus A350 and Boeing 787.

Why in this context has no HALE solar drone been able to demonstrate its ability to remain in the air for an unlimited period of time?

### 9.4.3. *A technological challenge: the aeroelasticity of flexible wings*

Whatever the current and expected progress of photovoltaic cells, the answer must be sought in all the disciplines involved in the design of an aircraft if one day we are to achieve unlimited endurance. The paradox mentioned above lies in the fact that the optimization of the whole object is not equivalent to the optimization of its parts. Indeed, if our solution is optimized from the point of view of aerodynamics, on the one hand, and the design of the structure, on the other hand, it is not very efficient at the border between the two. This boundary, called aeroelasticity, studies the mechanisms of interaction between the aerodynamic forces applied to the aircraft and the elasticity of its structure.

Wings with a high aspect ratio and very lightweight are typically very flexible, and this flexibility is a factor that can contribute to destructive interactions, of which the most well known is “flutter” (i.e. flapping of wings). This flutter phenomenon is both well known to the aviation community and a source of many misguided explanations. A precise explanation of the phenomenon is given in Box 9.1 with an overview in Box 9.2. These phenomena of interactions, which were under control at the end of the 20th Century, are returning to the forefront of the aeronautics sector with the exploration of endurance limits. This phenomenon was implicated in the accidents of NASA's Helios UAV and Facebook's Aquila UAV (part 3). The accident of the Google Solara 50 drone is the result of a similar phenomenon because it was linked to the flexibility of the wing.

**What is flutter?**

In practice, we refer to flutter when an aircraft's lifting surface (wing or tail) will, under certain flight conditions, undergo an oscillating movement coupled with bending (wing flapping) and twisting. This movement is amplified very rapidly, leading at best to a stabilized movement of high amplitude and at worst to the sudden destruction of the aircraft. The onset of this instability is associated with excessive speed and is one of the factors that determines the flight speed not to be exceeded for a given aircraft.

This phenomenon is often described as resonance, but this is not the case here; let us demonstrate this. In mechanics, we speak of resonance when a periodic input excitation is significantly amplified by a structure. There are many examples, such as a troop walking on a bridge that may cause the bridge to resonate if the step frequency corresponds to the bridge's natural frequency. Therefore, resonance means periodic excitation, an element absent from the flutter phenomenon: no need for wind gusts at a given frequency to trigger such a phenomenon; it is sufficient for a given altitude and Mach number to exceed a certain flight speed.

After seeing what it is not fluttering, let us unravel the mechanism that leads to its appearance. As mentioned above, this is an interaction between airflow and structural elasticity, an interaction on two aspects.

The first interaction concerns the natural oscillation frequencies of the wing. On the ground, just as a tuning fork excited by a shock will oscillate at a frequency of 440 Hz, a wing excited in flexion will beat at a frequency of a few Hz, while a torsional excitation triggers a pivoting oscillation of a higher frequency. In flight, the interaction with the flow will modify the value of these frequencies, increasing the first one and decreasing the second one to achieve the equality of both, a condition prone to instability.

This first interaction therefore allows the second to come into play. To understand it, it should first be noted that the lift force is directed upwards when the front of the wing (the leading edge) is directed upwards and downwards when the leading edge is directed downwards. In our case, the phasing between the flapping of the wing and its torsion leads to orient lift in the direction of movement. The leading edge is twisted up when the wing bends upwards and twisted down when the wing bends downwards. Under these conditions, the lift indefinitely increases the wing flapping until very large deformations, often to the point of failure, are obtained. This is a phenomenon known as dynamic instability by frequency coupling.

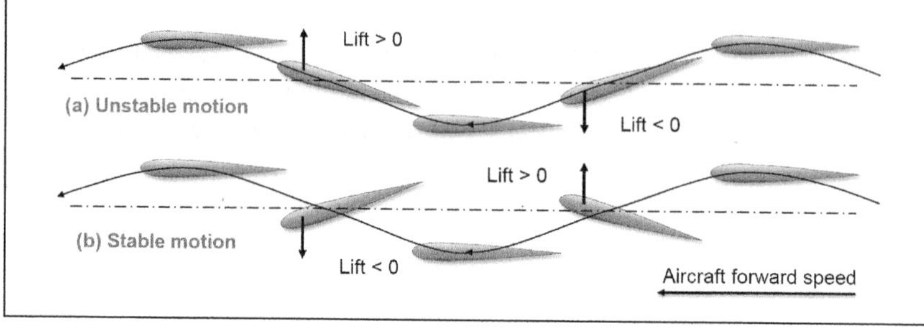


**Box 9.1.** *Explanation of the flutter phenomenon*

A brief history of flutter

The first case of flutter in the history of aeronautics was documented by F.W. Lanchester in 1916 [LAN 16] and concerned the tailplane of the Handley Page 0/400 biplane bomber. The interwar period was a very active period for this discipline due to the increase in the flight speed of aircraft at a faster rate than that of progress in the field of structural rigidity. Several dramatic cases were encountered in the 1920s and 1930s during Air Racers competitions such as the Curtis R6 in 1924 or the Gee Bee in 1931 [GAR 81]. The type of instabilities encountered after World War II evolved, and more problems were related to Mach effects occurring near the sound barrier as well as interactions with external loads under the wings (weapons, tanks, etc.). NACA, NASA's predecessor, identified 54 flutter-related incidents involving American military aircraft over the period 1947–1956. In France, a dramatic accident involving a horizontal tailplane flutter occurred during a test flight of the Mirage F1 fighter on May 18, 1967, which resulted in the death of pilot René Bigand.

**Box 9.2.** *A brief history of the flutter phenomenon*

These three accidents and the failures of these programs highlight the difficulty of implementing such devices. The low energy available forces designers to create a very large and very light wing and therefore necessarily very flexible. These devices cannot be designed by optimizing each discipline separately, i.e. aerodynamics, structure and flight dynamics, but require consideration of the physical phenomena that can result from the mixing of these disciplines. However, these are still not well known when the wings are very flexible, and many simulations have been in progress throughout the world since the early 2000s [SHE 07; PAL 10; KIR 17].

### 9.4.4. *An alternative way to remedy flutter: aeroelastic weaving*

Solutions to delay the critical rates of occurrence of these dangerous and even fatal phenomena were quickly found by aeronautical engineers. They are based on the modification of two main parameters, namely, weight distribution, on the one hand, and structural stiffening, on the other hand. The idea, in both cases, is to modify the natural frequencies of flexion and torsion of the wing in flight (Box 9.1).

The common point of these two solutions is to operate at the expense of the total weight of the aircraft and therefore ultimately at the expense of its endurance. It therefore appears necessary to develop alternative solutions to design maximum endurance aircraft in order to ensure a flight speed range compatible with the field of application that is free of instability and without using methods that overload the structure. Particular effort is also needed in the field of modeling these interaction phenomena due to the complexity of the instability mechanisms recently encountered.

Among these alternative solutions, we can mention *aeroelastic tailoring*, a technique that consists of exploiting the specificities of composite materials. These materials, known as laminates, are composed of successive plies of glass or carbon fibers oriented in particular directions, all immersed in a polymer resin, most often epoxy resin. Depending on the successive orientation of these fiber plies, couplings between these different deformation modes can emerge in the final material.

When applied to a wing, the bending deformation (wing flapping) can be coupled to the torsional deformation. This is a way of counteracting the natural tendency of the wing's flapping and torsional frequencies to meet as flight speed increases, thereby delaying the critical speeds at which flutter occurs.

This technique has two major advantages. First of all, it has little or no impact on the mass balance of the final airframe. It is a so-called "passive" technique in the sense that it does not use a mechanical or electrical actuator and thus removes the risk of failure. This technique was introduced in the 1970s on the American forward swept wing experimental aircraft X29 [PUT 84] (Figure 9.9). Its transposition to the case of high aspect ratio flexible wings of solar drones at high altitude is currently under investigation. It is complex because, as mentioned in the accident report of the Helios UAV [NOL 04], the physics at stake is characterized by strong coupling between the different disciplines of aerodynamics, structural mechanics, flight mechanics and automatic flight control laws.

The study of these complex phenomena is therefore by nature highly transdisciplinary and is still inaccessible to high-fidelity numerical simulation tools due to a prohibitive cost in computing time. A significant research effort is focused on the development of relevant simplified models capable of both modeling the couplings generated by aeroelastic tailoring and accurately predicting the rates of occurrence of instabilities such as flutter [KIR 17].

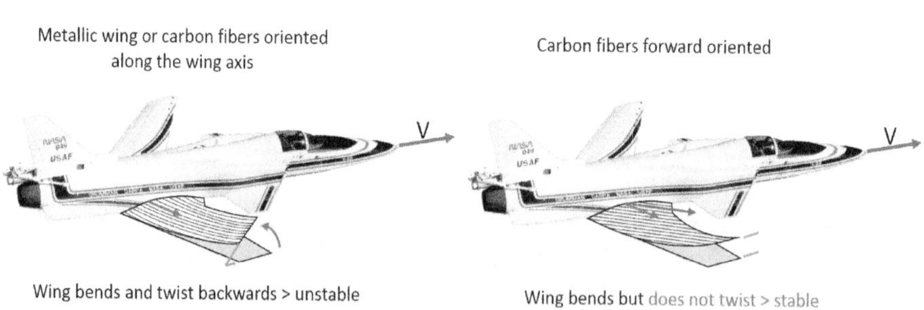


**Figure 9.9.** *Explanation of the principle of aeroelastic tailoring on the wing of the NASA X-29*

These research areas go beyond the restrictive framework of high-altitude solar drones since they represent a way of optimizing commercial aeronautics. The idea, supported in particular through the European 2020 Flexop program [FLE 15], is to design larger wings without increasing the mass balance and therefore more flexible while guaranteeing a flight domain free of aeroelastic instabilities. Aeroelastic tailoring is an example of a technological block being developed to achieve this objective; there are others such as active methods using piezoelectric materials to control the deformation of lifting surfaces.

## 9.5. Conclusion

Whether through a fixed-wing solar drone solution or one that is lighter than air, the endurance race is a powerful engine for research and innovation, particularly in the field of aeroelasticity, with recent accidents serving as a reminder of the road ahead. This scientific and technological challenge has been presented here through the prism of two complementary approaches: transdisciplinarity and the study of innovative solutions. These are the basis of technological building blocks which, when added together, will eventually lead to the realization of a concept born in the 1950s with the discovery of photovoltaic cells.

The case of high-altitude pseudo-satellites presented in this chapter perfectly illustrates the emulation and synergies that are created around an innovative concept between scientific research and the ecosystem capable of *ultimately* creating value [BAR 16]. Thus, the first successes obtained, such as the 14-day flight of the Zephyr 7 UAV, foreshadow the conquest of an almost vacant capability field located between conventional aircraft and satellites. The development of these innovative platforms should eventually make it possible to provide adaptable and reactive solutions for civil and military applications, both in the field of observation and telecommunications.

## 9.6. References

[BAR 16] BARBAROUX P., ATTOUR A., SCHENK E., *Knowledge Management and Innovation: Interaction, Collaboration, Openness*, ISTE Ltd, London and John Wiley & Sons, New York, 2016.

[BAU 15] BAURREAU F., STARAJ R., FERRERO F., LIZZI L., RIBERO J.M., CHESSEL J.P., "Stratospheric platform for telecommunication missions", in *Antennas and Propagation & USNC/URSI National Radio Science Meeting, IEEE International Symposium*, pp. 914–915, 2015.

[BOU 85] BOUCHER R.J., "Sunrise, the world's first solar-powered airplane", *Journal of Aircraft*, vol. 22, pp. 840–846, 1985.

[CES 06] CESTINO E., "Design of solar high altitude long endurance aircraft for multi payload & operations", *Aerospace Science and Technology*, vol. 10, no. 6, pp. 541–550, 2006.

[GAR 81] GARRICK I.E., REED W.H., "Historical development of aircraft flutter", *Journal of Aircraft*, vol. 18, no. 11, pp. 897–912, 1981.

[FLE 17] "FLEXOP". [Online]. Available at: https://www.flexop.eu/index.html. [Accessed 20 November 2017].

[FLI 98] FLITTIE K., CURTIN B., "Pathfinder solar-powered aircraft flight performance", in *23rd Atmospheric Flight Mechanics Conference*, paper 4446, 1998.

[HER 02] HERWITZ S., JOHNSON L., ARVESEN J., HIGGINS R., LEUNG J., DUNAGAN S., "Precision agriculture as a commercial application for solar-powered unmanned aerial vehicles", in *1st AIAA UAV Conference*, Portsmouth, paper 3404, 2002.

[JUV 16] JUVÉ L., FOSSE J., JOUBERT E., FOUQUET N., "Airbus group electrical aircraft program, the E-FAN project", in *52nd AIAA/SAE/ASEE Joint Propulsion Conference*, p. 4613, 2016.

[KIR 17] KIRSCH B., MONTAGNIER O., BENARD E., FAURE T., "Effet de l'anisotropie des composites sur le comportement aéroélastique des drones HALE", in *Journées Nationales sur les Composites 2017*, Champs-sur-Marne, 2017.

[LAN 16] LANCHESTER F.W., "Torsional vibrations of the tail of an aeroplane", *Aeronaut Res. Com R M*, vol. 276, 1916.

[MCC 83] MACCREADY P.B., LISSAMAN P.B.S., MORGAN W.R., BURKE J.D., "Sun-powered aircraft designs", *Journal of Aircraft*, vol. 20, no. 6, pp. 487–493, 1983.

[MON 09] MONTAGNIER O., BOVET L., "Optimisation of a solar-powered high altitude long endurance uav with composite wings", in *Proceedings of the 3rd EUropean Conference for AeroSpace Sciences*, 2009.

[MON 10] MONTAGNIER O., BOVET L., "Optimization of solar powered HALE UAV", in *Proceeding 2nd International Congress of the Aeronautics Science, ICAS 2010*, Cancun, 2010.

[MON 13] MONTAGNIER O., "Drones solaires : la quête du vol perpétuel", in MAZOYER S., DE LESPINOIS J., GOFFI E., BOUTHERIN G., PAJON C. (eds), *Les drones aériens: passé, présent et avenir. Approche globale*, La Documentation française, Paris, pp. 491–501, 2013.

[NOL 04] NOLL T.E., BROWN J.M., PEREZ-DAVIS M.E., ISHMAEL S.D., TIFFANY G.C., GAIER M., Investigation of the Helios prototype aircraft mishap, NASA Report, vol. 9, 2004.

[NOT 08] NOTH A., "Histoire de l'Aviation Solaire", *Laboratoire des Systèmes Autonomes*, ETH Zürich, Switzerland, 2008.

[NTS 17a] NTSB report Aquila. [Online]. Available at: https://www.ntsb.gov/_layouts/ntsb.aviation/brief.aspx?ev_id=20160701X62525&key=1. [Accessed 20 November 2017].

[NTS 17b] NTSB report Solara 50 [Online]. Available at: https://www.ntsb.gov/_layouts/ntsb.aviation/brief.aspx?ev_id=20150505X85410. [Accessed 20 November 2017].

[PAL 10] PALACIOS R., MURUA J., COOK R., "Structural and aerodynamic models in nonlinear flight dynamics of very flexible aircraft", *AIAA Journal*, vol. 48, no. 11, pp. 2648–2659, 2010.

[PUT 84] PUTNAM T.W., "X-29 flight-research program", 1984.

[ROM 13] ROMEO G., BORELLO F., CORREA G., CESTINO E., "ENFICA-FC: Design of transport aircraft powered by fuel cell & flight test of zero emission 2-seater aircraft powered by fuel cells fueled by hydrogen", *International Journal of Hydrogen Energy*, vol. 38, no. 1, pp. 469–479, 2013.

[SHE 07] SHEARER C., CESNIK C., "Non-linear flight dynamics of very flexible aircraft", *Journal of Aircraft*, vol. 44, no. 5, pp. 1528–1545, 2007.

# Conclusion

# Disruptive Technology and Defence Innovation Ecosystems: The Need for Dynamic Capabilities

## C.1. Overall conclusion

The changes brought about by the digital revolution, considered as a disruptive innovation [BOW 95], have attracted particular attention from public and private defence stakeholders. The objective of governments and defence companies is to renew military capabilities through the integration of new digital technologies into existing networks and weapon systems. In the long term, some analysts predict that the architecture of future combat networks will be composed of intelligent and autonomous systems, capable of collaborating with human agents to perform complex tasks. All application areas of military capabilities are concerned. Sensors, command and control systems, vectors, effectors, as well as logistics activities, military support and training are already affected by the development of robotics, artificial intelligence, computer simulation, nanotechnologies or neuroscience. In this context, the risk for traditional defence stakeholders is to undergo change rather than drive it, by not accurately anticipating the impact of new technologies on business models, professional practices and concepts of operations. As recalled [ASS 06: 218], disruptive innovation refers to "*a radically new concept, process, or product that significantly transforms demand and needs expressed in existing markets and industries, disrupts*

---

Conclusion written by Pierre BARBAROUX.

*established players and creates new markets or business practices, resulting in a significant societal impact*". However, in order to be able to anticipate the potential impact and benefit from disruptive innovations, it is necessary to have certain capacities. In particular, mastering the knowledge bases from which disruptive innovations develop requires the ability to detect, integrate and transform them in order to create value for the defence sector [BAR 16].

As an illustration, the Man–Machine Teaming (MMT) project, launched and financed by the French Procurement Agency (DGA) and led by Dassault Aviation and Thales, seeks to address the above capability issue. The project aims to advance the capacities of companies and research laboratories within the French ecosystem involved in the exploration, use or production of disruptive technologies in the fields of artificial intelligence and autonomy. The design of the Future Combat Aircraft System involves identifying, assimilating and integrating a variety of technological blocks, currently isolated, which are estimated to produce intelligent human–machine interfaces whose use would no longer be limited to the execution of simple (most often automated) tasks, but would be based on collaborative work between operators and air systems (source: MMT website; https://man-machine-teaming.com/presentation-generale/).

The example of the MMT project suggests that the capabilities required to make sense of the scientific, technological, political and legal environments, in which disruptive innovations are likely to create value for organizations, are essential but difficult to acquire and develop. In addition, these dynamic capabilities [TEE 07] are based on specific cognitive and affective microfoundations (e.g. attention, reflexivity) that are likely to shape individual and collective representations within organizations [HOD 11]. When adapted to the characteristics of the environment, these representations provide organizations with valuable resources to seize business opportunities through systematic and continuous observation of scientific and technological advances, sometimes far from their core competencies or market positioning. [GAV 05] showed how POLAROID, faced with the transformations brought about by the digital revolution in its business model, failed because of its inability to develop a set of common representations shared between its different hierarchical levels. According to [GAV 05], the bankruptcy of the company is precisely linked to the heterogeneity of internal representations about the threats and business opportunities generated by new digital imaging technologies: intermediate managers, considered closer to the field of operations, were unable to share with the company's managers their perceptions of the transformational opportunities that the new technology offered. Despite the mastery of advanced technical skills in the scientific and technical fields of digital imaging, the distance of the company's managers from day-to-day operations has reduced their ability to interpret information from operational experience and led to particularly disastrous strategic decisions. This effect, [GAV 05: 613] argued, "*is particularly strong when companies are engaged in multiple, heterogeneous domains, for which corporate managers, unlike divisional managers, have to access different, potentially contrasting action-outcome relationships*". [GAV 05] pointed to a link between the organization's structure and its ability to build shared representations based on the interpretation of a variety of information sources. Hence, in large multi-divisional hierarchical companies (e.g. armies, major defence groups), operational perceptions and experiences compete for the limited attention of high-level decision-makers, who systematically favor signals that are in line with the predictions of existing mental models, to the detriment of discordant and/or ambiguous feedback. This asymmetry is the basis of the *novelty paradox*: the new fields of action which, logically, should be the subject of deliberate attention in leaders, are most often associated with inappropriate representations, beliefs and mental models.

Another consequence of the variety of sources of disruption and their distance from the core competencies of organizations is the need to develop the capacity to absorb multiple sources of innovation, both internal and external [COH 90]. These can be broken down into four categories of cognitive resources [ZAH 02]: the acquisition, assimilation, transformation and exploitation of knowledge. These cognitive resources are encoded in organizational routines and mediated by the skills of the individuals and teams that make up the organizations. They also maintain mutually reinforcing relationships, with the acquisition and assimilation of new knowledge positively linked to each other, as well as the transformation and exploitation of knowledge [DAS 13]. [DIS 18] also suggested that interindividual differences (i.e. different cognitive profiles and a variety of individual skills; *knowledge workers*) contribute to the observed heterogeneity of firms' absorption capacities. Some individuals thus play a crucial role in the formation of organizational capacities, thus opening a research agenda on the selection and recruitment of individuals in charge of planning, conducting and supervising creative and/or knowledge-intensive activities (e.g. R&D, engineering, innovative project management, design). Research has also showed that the absorption capacity of firms is a matter of finding a balance between cognitive specialization and variety. This balance is considered favorable to the combination of heterogeneous and distributed knowledge and to the mobilization of exploration and exploitation-learning processes [LIC 09].

The defence community, faced with the need to benefit from, and even support, the development of disruptive innovations, faces a major challenge: to continuously develop the dynamic and absorptive capacities of the individuals and organizations that compose it. As recalled [ASS 06], there are many obstacles to the development of disruptive innovations in companies and, more broadly, within business ecosystems. Among these barriers, the organizational rigidities inherited from the past, the inability to "unlearn" certain established practices, risk aversion, the maintenance of inappropriate mental models, the weight of bureaucracy and the absence of distinctive skills in collective learning and the promotion of creativity, strongly influence the ability of companies and public organizations to benefit from disruptive innovations. This book has tried to show how defence stakeholders are meeting this challenge by adapting their knowledge bases, learning processes and organizational capacities in a turbulent and uncertain environment, through the experimentation of new models for managing disruptive innovation. In this respect, the proposal made by the President of the French Republic, Mr Emmanuel Macron, supported by the German Chancellor, Ms Angela Merkel, to create a Joint European Disruptive Initiative (JEDI) marks a decisive turning point in the development of the European technological ecosystem, both civil and military. The initiative now includes more than 150 European leaders from research centers and universities, technological start-ups and large companies, private and public investment companies, united around a mission: to identify, support and finance European technological advances that can promote a carbon-free world, a digital transition at the service of human beings, health for all and the exploration of new borders. MMT Project, Defence Innovation Agency and JEDI are three initiatives that highlight the need to promote disruptive innovation in the service of the French and European civilian and defence technological base and industrial capabilities.

## C.2. References

[BAR 16] BARBAROUX, P., ATTOUR, A., SCHENK, E. (2016), *Knowledge Management and Innovation: Interaction, Collaboration, Openness*, ISTE Ltd, London and John Wiley and Sons, New York.

[BOW 95] BOWER, J., CHRISTENSEN, C. (1995), "Disruptive technologies: Catching the waves", *Harvard Business Review*, January-February, pp. 43–53.

[COH 90] COHEN, W., LEVINTHAL, D. (1990), "Absorptive capacity: New perspective on learning and innovation", *Administrative Science Quarterly*, vol. 35, pp. 128–152.

[DAS 13] DASPIT, J., D'SOUZA, D. (2013), "Understanding the multi-dimensional nature of absorptive capacity", *Journal of Management Studies*, vol. 25, no. 3, pp. 299–316.

[DIS 18] DISTEL, A. (2018), "Unveiling the micro foundations of absorptive capacity: A study of Coleman's bathtub model", *Journal of Management*, [Online] Available at: https://journals.sagepub.com/doi/abs/10.1177/0149206317741963.

[GAV 05] GAVETTI, G. (2005), "Cognition and hierarchy: Rethinking the micro foundations of capabilities' development", *Organization Science*, vol. 16, no. 6, pp. 599–617.

[HOD 11] HODGKINSON, G.P., HEALY, M.P. (2011), "Psychological foundations of dynamic capabilities: Reflexion and reflection in strategic management", *Strategic Management Journal*, vol. 32, no. 13, pp. 1500–1516.

[LIC 09] LICHTENTHALER, U. (2009), "Absorptive capacity, environmental turbulence, and the complementarity of organizational learning processes", *The Academy of Management Journal*, vol. 52, no. 4, pp. 822–846.

[TEE 07] TEECE, D. (2007), "Explicating dynamic capabilities: The nature and micro foundations of (sustainable) enterprise performance", *Strategic Management Journal*, vol. 28, no. 13, pp. 1319–1350.

[ZAH 02] ZAHRA, S., GEORGE, G. (2002), "Absorptive capacity: A review, reconceptualization, and extension", *Academy of Management Review*, vol. 27, pp. 185–203.

# List of Authors

Walter ARNAUD
Direction Générale de l'Armement
Paris
France

Pierre BARBAROUX
Ecole de l'Air
Chaire Cyber Résilience
Aérospatiale,
Armée de l'Air
Salon-de-Provence
France

Jean BELIN
GREThA (UMR 5113)
University of Bordeaux
Chaire Economie de défense
IHEDN – Ecole Militaire
Paris
France

Cyril CAMACHON
Ecole de l'Air
Salon-de-Provence
France

Cécile FAUCONNET
Université Paris 1 Panthéon-Sorbonne
ENSTA ParisTech – Unité d'économie appliquée (UEA)
Paris
France

Vincent FERRARI
CReA, Ecole de l'Air
Salon-de-Provence
France

Christophe GRANDEMANGE
Direction Générale de l'Armement
Paris
France

Marianne GUILLE
LEMMA (EA 4442)
LEMMA & LabEx MME-DII
Université Paris II Panthéon-Assas
Paris
France

Nicolas HUÉ
Direction Générale de l'Armement
Paris
France

Bertrand KIRSCH
Ecole de l'Air
Salon-de-Provence
France

Didier LEBERT
ENSTA ParisTech – Unité d'économie appliquée (UEA)
Paris
France

François-Xavier MEUNIER
ENSTA ParisTech – Unité d'économie appliquée (UEA)
Paris
France

Olivier MONTAGNIER
Ecole de l'Air
Salon-de-Provence
France

Sylvain MOURA
Observatoire Economique de la Défense
Paris
France

Célia ZYLA
SATT Lutech
Paris
France

# Index

Other titles from

in

Innovation, Entrepreneurship and Management

**2019**

AMENDOLA Mario, GAFFARD Jean-Luc
*Disorder and Public Concern Around Globalization*

DOU Henri, JUILLET Alain, CLERC Philippe
*Strategic Intelligence for the Future 1: A New Strategic and Operational Approach*
*Strategic Intelligence for the Future 2: A New Information Function Approach*

FRIMOUSSE Soufyane
*Innovation and Agility in the Digital Age*
*(Human Resources Management Set – Volume 2)*

HELLER David, DE CHADIRAC Sylvain, HALAOUI Lana, JOUVET Camille
*The Emergence of Start-ups*
*(Economic Growth Set – Volume 1)*

HÉRAUD Jean-Alain, KERR Fiona, BURGER-HELMCHEN Thierry
*Creative Management of Complex Systems*
*(Smart Innovation Set – Volume 19)*

LATOUCHE Pascal
*Open Innovation: Corporate Incubator*
*(Innovation and Technology Set – Volume 7)*

LEHMANN Paul-Jacques
*The Future of the Euro Currency*

LEIGNEL Jean-Louis, MÉNAGER Emmanuel, YABLONSKY Serge
*Sustainable Enterprise Performance: A Comprehensive Evaluation Method*

MILLOT Michel
*Embarrassment of Product Choices 2: Towards a Society of Well-being*

N'GOALA Gilles, PEZ-PÉRARD Virginie, PRIM-ALLAZ Isabelle
*Augmented Customer Strategy: CRM in the Digital Age*

NIKOLOVA Blagovesta
*The RRI Challenge: Responsibilization in a State of Tension with Market Regulation*
*(Innovation and Responsibility Set – Volume 3)*

PRIOLON Joël
*Financial Markets for Commodities*

QUINIOU Matthieu
*Blockchain: The Advent of Disintermediation*

ROGER Alain, VINOT Didier
*Skills Management: New Applications, New Questions*
*(Human Resources Management Set – Volume 1)*

SERVAJEAN-HILST Romaric
*Co-innovation Dynamics: The Management of Client-Supplier Interactions for Open Innovation*
*(Smart Innovation Set – Volume 20)*

SKIADAS Christos H., BOZEMAN James R.
*Data Analysis and Applications 1: Clustering and Regression, Modeling-estimating, Forecasting and Data Mining*
*(Big Data, Artificial Intelligence and Data Analysis Set – Volume 2)*
*Data Analysis and Applications 2: Utilization of Results in Europe and Other Topics*
*(Big Data, Artificial Intelligence and Data Analysis Set – Volume 3)*

## 2018

BURKHARDT Kirsten
*Private Equity Firms: Their Role in the Formation of Strategic Alliances*

CALLENS Stéphane
*Creative Globalization*
*(Smart Innovation Set – Volume 16)*

CASADELLA Vanessa
*Innovation Systems in Emerging Economies: MINT – Mexico, Indonesia, Nigeria, Turkey*
*(Smart Innovation Set – Volume 18)*

CHOUTEAU Marianne, FOREST Joëlle, NGUYEN Céline
*Science, Technology and Innovation Culture*
*(Innovation in Engineering and Technology Set – Volume 3)*

CORLOSQUET-HABART Marine, JANSSEN Jacques
*Big Data for Insurance Companies*
*(Big Data, Artificial Intelligence and Data Analysis Set – Volume 1)*

CROS Françoise
*Innovation and Society*
*(Smart Innovation Set – Volume 15)*

DEBREF Romain
*Environmental Innovation and Ecodesign: Certainties and Controversies*
*(Smart Innovation Set – Volume 17)*

DOMINGUEZ Noémie
*SME Internationalization Strategies: Innovation to Conquer New Markets*

ERMINE Jean-Louis
*Knowledge Management: The Creative Loop*
*(Innovation and Technology Set – Volume 5)*

GILBERT Patrick, BOBADILLA Natalia, GASTALDI Lise,
LE BOULAIRE Martine, LELEBINA Olga
*Innovation, Research and Development Management*

IBRAHIMI Mohammed
*Mergers & Acquisitions: Theory, Strategy, Finance*

LEMAÎTRE Denis
*Training Engineers for Innovation*

LÉVY Aldo, BEN BOUHENI Faten, AMMI Chantal
*Financial Management: USGAAP and IFRS Standards*
*(Innovation and Technology Set – Volume 6)*

MILLOT Michel
*Embarrassment of Product Choices 1: How to Consume Differently*

PANSERA Mario, OWEN Richard
*Innovation and Development: The Politics at the Bottom of the Pyramid*
*(Innovation and Responsibility Set – Volume 2)*

RICHEZ Yves
*Corporate Talent Detection and Development*

SACHETTI Philippe, ZUPPINGER Thibaud
*New Technologies and Branding*
*(Innovation and Technology Set – Volume 4)*

SAMIER Henri
*Intuition, Creativity, Innovation*

TEMPLE Ludovic, COMPAORÉ SAWADOGO Eveline M.F.W.
*Innovation Processes in Agro-Ecological Transitions in Developing Countries*
*(Innovation in Engineering and Technology Set – Volume 2)*

UZUNIDIS Dimitri
*Collective Innovation Processes: Principles and Practices*
*(Innovation in Engineering and Technology Set – Volume 4)*

VAN HOOREBEKE Delphine
*The Management of Living Beings or Emo-management*

## 2017

AÏT-EL-HADJ Smaïl
*The Ongoing Technological System*
*(Smart Innovation Set – Volume 11)*

BAUDRY Marc, DUMONT Béatrice
*Patents: Prompting or Restricting Innovation?*
*(Smart Innovation Set – Volume 12)*

BÉRARD Céline, TEYSSIER Christine
*Risk Management: Lever for SME Development and Stakeholder Value Creation*

CHALENÇON Ludivine
*Location Strategies and Value Creation of International Mergers and Acquisitions*

CHAUVEL Danièle, BORZILLO Stefano
*The Innovative Company: An Ill-defined Object*
*(Innovation Between Risk and Reward Set – Volume 1)*

CORSI Patrick
*Going Past Limits To Growth*

D'ANDRIA Aude, GABARRET Inés
*Building 21st Century Entrepreneurship*
*(Innovation and Technology Set – Volume 2)*

DAIDJ Nabyla
*Cooperation, Coopetition and Innovation*
*(Innovation and Technology Set – Volume 3)*

FERNEZ-WALCH Sandrine
*The Multiple Facets of Innovation Project Management*
*(Innovation between Risk and Reward Set – Volume 4)*

FOREST Joëlle
*Creative Rationality and Innovation*
*(Smart Innovation Set – Volume 14)*

GUILHON Bernard
*Innovation and Production Ecosystems*
*(Innovation between Risk and Reward Set – Volume 2)*

HAMMOUDI Abdelhakim, DAIDJ Nabyla
*Game Theory Approach to Managerial Strategies and Value Creation*
*(Diverse and Global Perspectives on Value Creation Set – Volume 3)*

LALLEMENT Rémi
*Intellectual Property and Innovation Protection: New Practices and New Policy Issues*
*(Innovation between Risk and Reward Set – Volume 3)*

LAPERCHE Blandine
*Enterprise Knowledge Capital*
*(Smart Innovation Set – Volume 13)*

LEBERT Didier, EL YOUNSI Hafida
*International Specialization Dynamics*
*(Smart Innovation Set – Volume 9)*

MAESSCHALCK Marc
*Reflexive Governance for Research and Innovative Knowledge*
*(Responsible Research and Innovation Set – Volume 6)*

MASSOTTE Pierre
*Ethics in Social Networking and Business 1: Theory, Practice and Current Recommendations*
*Ethics in Social Networking and Business 2: The Future and Changing Paradigms*

MASSOTTE Pierre, CORSI Patrick
*Smart Decisions in Complex Systems*

MEDINA Mercedes, HERRERO Mónica, URGELLÉS Alicia
*Current and Emerging Issues in the Audiovisual Industry*
*(Diverse and Global Perspectives on Value Creation Set – Volume 1)*

MICHAUD Thomas
*Innovation, Between Science and Science Fiction*
*(Smart Innovation Set – Volume 10)*

PELLÉ Sophie
*Business, Innovation and Responsibility*
*(Responsible Research and Innovation Set – Volume 7)*

SAVIGNAC Emmanuelle
*The Gamification of Work: The Use of Games in the Workplace*

SUGAHARA Satoshi, DAIDJ Nabyla, USHIO Sumitaka
*Value Creation in Management Accounting and Strategic Management: An Integrated Approach*
*(Diverse and Global Perspectives on Value Creation Set –Volume 2)*

UZUNIDIS Dimitri, SAULAIS Pierre
*Innovation Engines: Entrepreneurs and Enterprises in a Turbulent World*
*(Innovation in Engineering and Technology Set – Volume 1)*

## 2016

BARBAROUX Pierre, ATTOUR Amel, SCHENK Eric
*Knowledge Management and Innovation*
*(Smart Innovation Set – Volume 6)*

BEN BOUHENI Faten, AMMI Chantal, LEVY Aldo
*Banking Governance, Performance And Risk-Taking: Conventional Banks Vs Islamic Banks*

BOUTILLIER Sophie, CARRÉ Denis, LEVRATTO Nadine
*Entrepreneurial Ecosystems (Smart Innovation Set – Volume 2)*

BOUTILLIER Sophie, UZUNIDIS Dimitri
*The Entrepreneur (Smart Innovation Set – Volume 8)*

BOUVARD Patricia, SUZANNE Hervé
*Collective Intelligence Development in Business*

GALLAUD Delphine, LAPERCHE Blandine
*Circular Economy, Industrial Ecology and Short Supply Chains*
*(Smart Innovation Set – Volume 4)*

GUERRIER Claudine
*Security and Privacy in the Digital Era*
*(Innovation and Technology Set – Volume 1)*

MEGHOUAR Hicham
*Corporate Takeover Targets*

MONINO Jean-Louis, SEDKAOUI Soraya
*Big Data, Open Data and Data Development*
*(Smart Innovation Set – Volume 3)*

MOREL Laure, LE ROUX Serge
*Fab Labs: Innovative User*
*(Smart Innovation Set – Volume 5)*

PICARD Fabienne, TANGUY Corinne
*Innovations and Techno-ecological Transition*
*(Smart Innovation Set – Volume 7)*

## 2015

CASADELLA Vanessa, LIU Zeting, DIMITRI Uzunidis
*Innovation Capabilities and Economic Development in Open Economies*
*(Smart Innovation Set – Volume 1)*

CORSI Patrick, MORIN Dominique
*Sequencing Apple's DNA*

CORSI Patrick, NEAU Erwan
*Innovation Capability Maturity Model*

FAIVRE-TAVIGNOT Bénédicte
*Social Business and Base of the Pyramid*

GODÉ Cécile
*Team Coordination in Extreme Environments*

MAILLARD Pierre
*Competitive Quality and Innovation*

MASSOTTE Pierre, CORSI Patrick
*Operationalizing Sustainability*

MASSOTTE Pierre, CORSI Patrick
*Sustainability Calling*

## 2014

DUBÉ Jean, LEGROS Diègo
*Spatial Econometrics Using Microdata*

LESCA Humbert, LESCA Nicolas
*Strategic Decisions and Weak Signals*

## 2013

HABART-CORLOSQUET Marine, JANSSEN Jacques, MANCA Raimondo
*VaR Methodology for Non-Gaussian Finance*

## 2012

DAL PONT Jean-Pierre
*Process Engineering and Industrial Management*

MAILLARD Pierre
*Competitive Quality Strategies*

POMEROL Jean-Charles
*Decision-Making and Action*

SZYLAR Christian
*UCITS Handbook*

## 2011

LESCA Nicolas
*Environmental Scanning and Sustainable Development*

LESCA Nicolas, LESCA Humbert
*Weak Signals for Strategic Intelligence: Anticipation Tool for Managers*

MERCIER-LAURENT Eunika
*Innovation Ecosystems*

## 2010

SZYLAR Christian
*Risk Management under UCITS III/IV*

## 2009

COHEN Corine
*Business Intelligence*

ZANINETTI Jean-Marc
*Sustainable Development in the USA*

## 2008

CORSI Patrick, DULIEU Mike
*The Marketing of Technology Intensive Products and Services*

DZEVER Sam, JAUSSAUD Jacques, ANDREOSSO Bernadette
*Evolving Corporate Structures and Cultures in Asia: Impact of Globalization*

## 2007

AMMI Chantal
*Global Consumer Behavior*

## 2006

BOUGHZALA Imed, ERMINE Jean-Louis
*Trends in Enterprise Knowledge Management*

CORSI Patrick *et al.*
*Innovation Engineering: the Power of Intangible Networks*

Printed and bound by CPI Group (UK) Ltd, Croydon, CR0 4YY
13/05/2022
03124177-0003